PRIMATES

Edited by **Mark Burke** and **Maurice Ptito**

Primates
http://dx.doi.org/10.5772/65832
Edited by Mark Burke and Maurice Ptito

Contributors

Daya Shankar Gupta, Silmar Teixeira, Colin Groves, Greg Simmons, Gervais Habarugira, Renato Sathler-Avelar, Armanda Moreira Mattoso-Barbosa, Olindo Assis Martins-Filho, Andrea Teixeira-Carvalho, Danielle Vitelli-Avelar, Milada Řeháková, Mark Burke, Maurice Ptito, Joseph Bouskila, Kathleen Rockland, Carole Parron, Patrick Waeber, Lena Reibelt

First published in London, United Kingdom, 2018 by IntechOpen
IntechOpen is the global imprint of INTECHOPEN LIMITED, registered in England and Wales, registration number: 11086078, The Shard, 25th floor, 32 London Bridge Street
London, SE19SG – United Kingdom
Printed in Croatia

British Library Cataloguing-in-Publication Data
A catalogue record for this book is available from the British Library

Additional hard copies can be obtained from orders@intechopen.com

Primates, Edited by Mark Burke and Maurice Ptito
p. cm.
Print ISBN 978-1-78923-216-5
Online ISBN 978-1-78923-217-2

For EU product safety concerns:
IN TECH d.o.o., Prolaz Marije Krucifikse Kozulić 3, 51000 Rijeka, Croatia,
info@intechopen.com or visit our website at intechopen.com.

Meet the editors

Dr. Mark Burke graduated from the State University of New York, Plattsburgh, with an undergraduate degree in Psychology. He then went on to complete his doctoral degree in Biology from the McGill University with a postdoctoral fellowship at the University of Montreal. He is currently working as an associate professor at the Howard University and is a member of the DC Center for AIDS Research (CFAR). His research focuses on neurodevelopment and the effects of developmental intrusions in nonhuman primates, such as hemispherectomies, fetal alcohol exposure, and pediatric HIV infection. His current work aims at identifying the extent and underlying the mechanisms of HIV-induced neurological damage in infants. He is the PI of an NIH R03 grant and has been an active mentor of undergraduate, graduate, and medical students.

Maurice Ptito is a professor of Visual Neuroscience at the School of Optometry at the Université de Montréal. He is also an adjunct professor of Neurology and Neurosurgery at the Montreal Neurological Institute (McGill University) and a guest professor at the Department of Neuroscience and Pharmacology at the University of Copenhagen. He currently holds the Harland Sanders Research Chair in Vision Science. He has received several awards and prizes, including the Sir John William Dawson Medal of the Royal Society of Canada and the Henry and Karla Hansen Prize (Denmark). He is a fellow of the Royal Society of Canada and the Royal Danish Academy of Sciences and Letters. In 2013, he was made a Knight of the National Order of Quebec. The major goal of his research is the understanding of the mechanisms involved in visual plasticity in normal developing individuals, as well as in those who were born without vision or lost their vision later in life.

Contents

Preface

For over a century, there has been a recognition that through careful study and observation of primates, critical insights into human evolution, psyche, and biology continue to be identified. At the beginning of the twentieth century, Eugene Marias (*My Friends the Baboons*) went as far as to imply that for the very existence and perseverance of humanity, we must learn from our closest related species, that is, primates. Although this is a broad statement, Marias argued that the fight against deadly diseases, surgical procedures, and understanding of our own self-awareness and psyche have all been made possible by primate research. He also recognized the importance of observing primates in their natural habitat as their adaptability to a multitude of ever-changing environments provides unique clues about human capabilities. Our understanding of primates is an ongoing process with the complexity and richness of primate life and interactions demonstrating an intraspecific variability that mirrors interactions, adaptability, and richness of human cultures and societies.

This book contains 10 chapters that address primate phylogeny, natural observations, primate ecosystem, and mechanisms to minimize the consequences of human activity on natural habitat, sociocognitive abilities, disease pathophysiology, and biomedical research. The first part of this book examines the phylogeny of primates as an ongoing process that is evolving as new genetic- and molecular-based investigations, which propels this field. The second part examines the natural observations of the Philippine tarsier (one of the least studied primates) and human-primate interactions in Madagascar. These chapters provide a unique insight into the labor-intensive naturalistic observations in a tropical setting and an innovative effort to work with local governments and communities to ensure preservation of habitat. The third part examines the primate sociocognitive abilities by examining neural-temporal units and information processing in time budget and self-recognition in primates. The fourth section involves examining the pathophysiology and origin of gibbon ape leukemia and Chagas disease. These studies highlight the contribution of primate research to the biomedical field as well as provide a platform to examine and investigate diseases that have the potential to harm humans. The 30-plus-year history of using macaques to study the pathophysiology of human immunodeficiency virus (HIV) and the recent discovery of potential neurodevelopmental impairment as a result of both pre- and postnatal Zika infections are a further testament to the power and continual need for primate research. Finally, the last section of this book examines species-specific neuroanatomical characteristics of primate brain through the investigation of the endocannabinoid system and white matter tracks.

Contributions to this volume originate from distinguished faculty from a multitude of countries including the United States, Denmark, Australia, Canada, Brazil, Czech Republic, France, Madagascar, and Switzerland. We are grateful for their thoughtful and stimulating chapters that will be useful for the graduate students, instructors, and researchers as a springboard for discussion and generation of research questions.

We express our gratitude to the Behavioural Sciences Foundation, St. Kitts, which has supported our research for over 20 years. I am personally grateful to my graduate supervisors, Roberta Palmour and Frank Ervin, who provided me with my first exposure to primate research as a young graduate student. During my first research trip to work with the monkeys, I was excited to dive-in to the experimental procedures, perhaps a bit too much as my graduate mentors were quick to tell me to slow down, appreciate, and understand the primate behavior. We are also indebted to InTechOpen Science for their effort and support for the publishing of this volume with a special appreciation to Ms. Romina Skomersic, Publishing Process Manager, for her tireless work on this edited volume.

Dr. Mark Burke
Department of Physiology & Biophysics
College of Medicine, Howard University
Washington, DC, USA

Dr. Maurice Ptito
Department of Physiology, School of Optometry
Université de Montréal
Montréal, QC, Canada

Chapter 1

Introductory Chapter: Primates - What the Monkey Brain Tells the Human Brain

Mark W. Burke and Maurice Ptito

Additional information is available at the end of the chapter

http://dx.doi.org/10.5772/intechopen.76482

1. Introduction: The Soul of The Ape

The name Eugene Marais has slowly begun to fade to the annals of time; however, we would be remiss to begin a book about Primates without first discussing the life and work of the man who helped lay the foundation for naturalistic primate observation. Born in 1871 outside of Pretoria in South Africa, he first started out as a journalist who built a reputation for upsetting politicians to the point where he was indicted for high treason. After his acquittal, he moved to London where he not only studied law but also had tried his hand at medicine. During his time in London, the Boer War had begun and Marais left for Central Africa where he attempted to help his countrymen. Early in the twentieth century, estimated to be around 1903, Marais retreated to Waterberg, a mountainous region in the north Limpopo Province of South Africa. The farmers who were originally in that area had largely been displaced as a result of the Boer War and because of this, the chacma baboons (*Papio ursinus*) had a temporary reprieve from human interaction. It was this time in Waterberg that Marais spent 3 years living with, following, and studying the chacma. He became one of the first to study wild baboons in their natural environment and consequently wrote "My Friends the Baboons" and his unfinished work "The Soul of the Ape." For the most part, Marais was an untrained scientist, except perhaps a brief medical introduction during his time in London, but this might have led to the strength of his investigation by not having preconceived notions tainting his observations [1].

In these two works, Marais delved into the psyche or as he termed it, the "soul" as he questioned phyletic (unconscious/instinctual) versus casual (conscious/learned) memory. Through time, he was able to make observations within a few yards of the chacma troop. However, as time passed, farmers along with their guns returned, thus finishing the relationship between Marais and the troop. Although this work remains unfinished, his insights about the psyche would not be possible without his keen observations of the chacma:

"The phyletic history of the primate soul can clearly be traced in the mental evolution of the human child. The highest primate, man, is born an instinctive animal…as it grows, the new mentality slowly, by infinite gradations, emerges…it is here that the wonderful transition occurs, a transition which the phyletic evolution of the soul of the chacma exemplifies. As the new soul, the soul of the individual memory slowly emerges, the instinctive soul becomes just as slowly submerged." [1], *pp. 102–103*

2. Primates and Intoxication

In addition to the psyche, Marais' observations and perhaps his own personal experiences, delved into addiction and depression as he stated:

"Euphoric intoxication is of especial interest in this study because of convincing proof that there exists in the chacma a state of mind similar to that which induces the use of euphoric in man." [1], *p. 117*

This is of special interest to our research group as vervet monkeys (*Chlorocebus sabeus*) will, in naturalistic settings, voluntarily consume alcohol [2, 3]. In fact, the St. Kitts vervet will drink beverage alcohol in both the laboratory and natural settings, with 15% voluntarily consuming over 5 g of ethanol/kg/day [2]. The range of alcohol consumption in this population is similar to that seen in the human population that varies from abstinence to those that chose to drink to the point of comatose with perhaps the largest population being somewhere in the middle. The consumption of alcohol has its roots in our evolutionary frugivorous history. The presence of ethanol in fruits coincides with ripeness and sugar content and with a potentially higher caloric content. As a result, it would have been advantageous to consume ripe fruit that has started to ferment [4]. Voluntary alcohol intake has been noted in different species including birds, baboons, elephants, and the aforementioned vervets [1, 4–6]. It has been hypothesized that the excessive consumption of alcohol is due to an advantageous ancestral trait that has become disadvantageous due to the abundant access of nutrition [4, 7]. This adaptive mechanism has been suggested to be related to "exploratory appetitive behavior" involving neurogenic as opposed to the neurological effects of ethanol [7]. The neurological effects of ethanol (sedative, tolerance, anxiolytics, and dependence) are important factors in the development and sustenance of alcohol abuse; it remains a complex disorder [2], as Marais would attribute this to both phyletic and casual memory of the species. Although the etiology of alcoholism is unknown in humans, the likelihood of alcoholism in nonhuman primates sharing some aspects of the same etiology is great. In fact, Marais recognized the overlap of the psyche between human and nonhuman primates a 100 years ago, stating:

"…the conclusion that the chacma suffers from the same attribute of pain which is such an important ingredient of human mentality, and that the condition is due to the same cause." [1], *p. 139*

With the voluntary and naturalistic drinking pattern exhibited by vervets, we are able to address vulnerability factors, both genetic and neurochemical, leading to alcohol use and misuse [2]. We have been able to now further take advantage of the drinking patterns to examine the short-and long-term effects of prenatal ethanol exposure on the developing brain in a systematic manner which cannot be done in a clinical setting [3, 8–10].

3. Validity of Model Systems

Given similarities such as neuroanatomy, physiology, immune, development, behavior, and anatomy and those outlined by Marais [1], between human and nonhuman primates, it might be tempting to restrict models of human conditions to monkeys. However, from both a practical and an ethical point of view, it is appropriate to use lower mammals to test initial hypothesis, while reserving the study of nonhuman primates to final confirmations of hypothesis already well piloted in lower mammals and to situations in which other model systems do not provide an adequate degree of complexity. Furthermore, the validity of any animal model, including nonhuman primates, depends on the question being asked. To evaluate animal model validity, five criteria consisting of *homological* (assess species and strain), *pathogenic* (disease process similarities), *mechanistic* (assess proposed mechanisms of action as it relates to the human condition), *face* (similarity of observable disease features), and *predictive* (ability of model to make predictions on therapeutic interventions) validity should be examined as it relates to the research question [11–13].

For instance, applying the test of validity to our longitudinal assessment of the functional reorganization and adaptive neuroplastic responses following early life hemispherectomies provides a demonstration of the strength of nonhuman primate model systems. The premise of developing this model was due in part to the remarkable recovery of patients following the cerebral hemicorticectomy surgical procedure as a treatment for intractable epilepsy [14–16]. The degree of recovery in the clinical setting depends on the age of intervention and the targeted sensory and motor system [15–17]. Following surgical intervention, there appears to be a rapid recovery of sensory and motor systems, which may be a result of a preexisting functional reorganization due to the dysfunctional hemisphere [18, 19]. The purpose of our nonhuman primate model is to model the functional recovery by identifying the resultant reorganization and behavioral recovery following infant and adult hemispherectomy [20].

Infant (aged about 9 weeks) and adult (about 48 months of age) vervets underwent a surgical procedure to remove the left cerebral hemisphere (**Figure 1**) and allowed to recover in enriched environments at the Behavioral Sciences Foundation, St. Kitts. Behavioral assessments were conducted on a 6-month basis. All surgical and behavioral protocols were approved by the Animal Care and Use Committee at University of Montreal.

Sensory assessments consisted of visual (perimetry, palpebral reflex, and visual pursuit), thermal, and nocioceptive tasks, while motor observations were conducted via open field and horizontal bar crossing (as reviewed in Burke et al. [20]). Infant hemispherectomized monkeys displayed residual vision in the "blind" hemifield up to 45°, but adult subjects were unable to detect visual stimuli. Normal-sighted monkeys had a visual perimetry up to 90° in both hemifields. For both the infant- and adult-lesioned subjects lacked visual palpebral reflex and visual pursuit in the contralateral visual field. Infant-lesioned subjects retained nocioceptive and innocuous sensation capabilities on the contralateral side. In the open field,

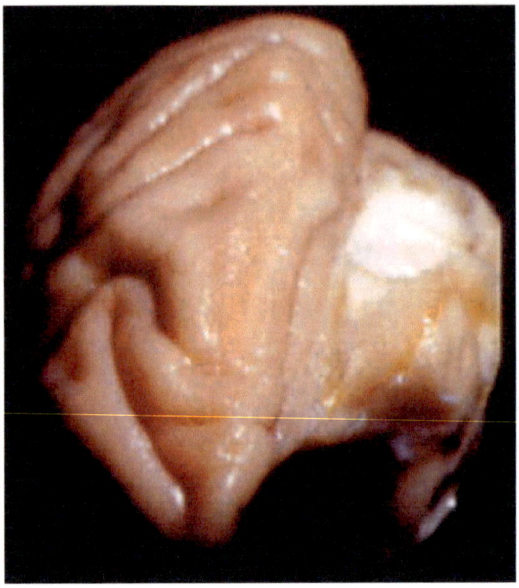

Figure 1. Hemispherectomized brain: image here shows the removal of the left hemisphere (adapted from Burke et al. [20]).

we observed normal ipsilateral upper and lower limb and contralateral lower limb gait. Upper limb on the contralateral side remained paretic in infant-lesioned subjects. This is in contrast to adult-lesioned subjects where both upper and lower limbs were paretic. Within the first 2 years after surgery, infants displayed difficulty traversing the horizontal bar, after which the infant-lesioned subjects were able to cross by walking upright, but like the open-field observations, subjects did not attempt to use the upper contralateral limb (**Figure 2**). Subjects also displayed ipsiversive and circling behaviors, possibly due to contralateral hemianopia [20].

Given the neurodevelopmental and homologous brain areas, the nonhuman primate offers a high degree of homological validity for the study of human development [3, 8, 9, 11, 20–24]. Pathogenic validity, which addresses the disease process, in this case, depends on the question. If, for instance, the question in this case were to model reorganization following hemispherectomy as a result of intractable epilepsy, then the model would have to also show a similar pathogenic process. However, this model is aimed at identifying the ability of the brain to functionally remodel during early development in a manner that cannot be fully elucidated in the human epileptic condition considering the potential for preexisting reorganization [20]. Several lines of evidence from our study demonstrate both *mechanistic* (addressing mechanism of action) and *face* (similarities of observable features) validity for neural remodeling manifested through behavior as well as histological alterations. The residual vision and pervasive ipsilateral turning seen in our subjects is reminiscent of hemianopia seen in hemispherectomized patients in which a subset has residual responses to visual stimuli on the hemianopic field, known as Type I blindsight (implicit) or Type II blindsight (explicit) [17]. The ipsilateral turning corresponds to the visual preference reported in the clinical population [25]. Anatomically, our subjects have a significant degeneration of foveal retinal ganglion

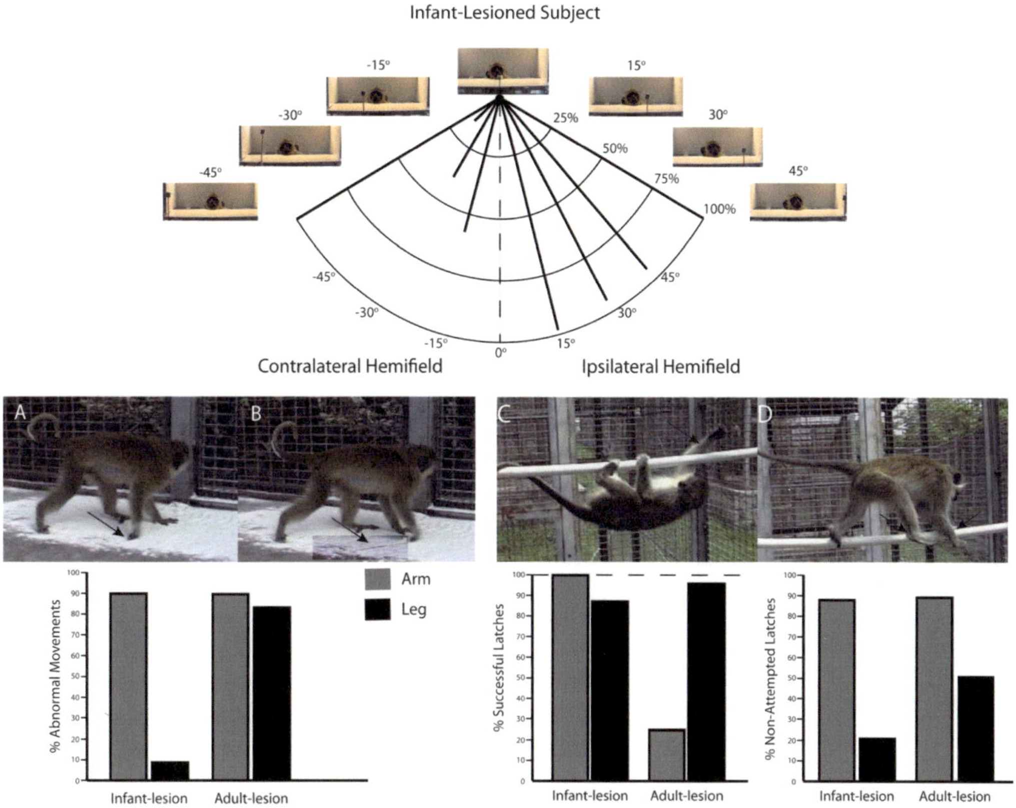

Figure 2. Behavioral analysis: The perimetry test (top image) depicts residual vision. Subjects were able to detect visual stimuli at 45° in the blind field at a 16% success rate (contralateral hemifield) with no responses elicited beyond 45°. Panels A and B depict the open-field test where normal gait was significantly affected by surgery. The contralateral upper limb in the infant-lesioned subjects displayed paresis; however, the lower limb showed little residual paresis. In the adult-lesioned subject, however, both upper and lower limbs showed significant paresis. Panel C shows a subject 1-year post surgery unable to cross the horizontal bar in an upright position, whereas in panel D at 2-year post surgery, the subject is able to cross in an upright position. Furthermore, in panel C, the young monkey is unable to grasp the bar with the upper limb and would glide the arm along the bar while attempting to grasp the bar. By 2–3 years' post surgery, the subject is able to have more successful latches per attempt, but for most of the trials, the subjects did not attempt to latch on with the upper limb. Graphical data are shown at a 3-year post-surgical time points for all groups (adapted from Burke et al. [20]).

cells but remain intact in the peripheral retina [26]. The ipsilateral lateral geniculate also suffers massive neural degeneration; however, it too retains neurons and appropriately placed projections from the retina despite significant volume loss [27, 28]. Likewise, human hemispherectomy patients regain strength in lower limbs but display significant weakness in the contralateral upper limb [19, 29]. Residual contralateral tactile sensation remains intact and activates the ipsilateral somatosensory cortices [30, 31]. Histological data from our monkeys suggest that the dorsal column nuclei (cuneate and gracilis subdivisions) are unaffected, providing an anatomical substrate for an intact ipsilateral, non-decussating pathway. The residual vision and more complete motor recovery in infant, but not adult-lesioned, subjects further supports clinical data of a profound functional reorganization of neural circuitry underlying the behavioral observations [20].

Predictive validity (ability to make predictions) typically examines pharmacological interventions. However, here, we propose a functional model for neuroanatomical reorganization in multiple systems. Clinical functional magnetic resonance imaging (fMRI) data suggest ipsilateral sensory and motor pathways [30–32] potentially through the corticospinal and medial lemniscus tracts [33–35]. Our model supports several lines of evidence from the visual pathway that allow us to propose a neuroanatomical substrate for residual vision (**Figure 3**),

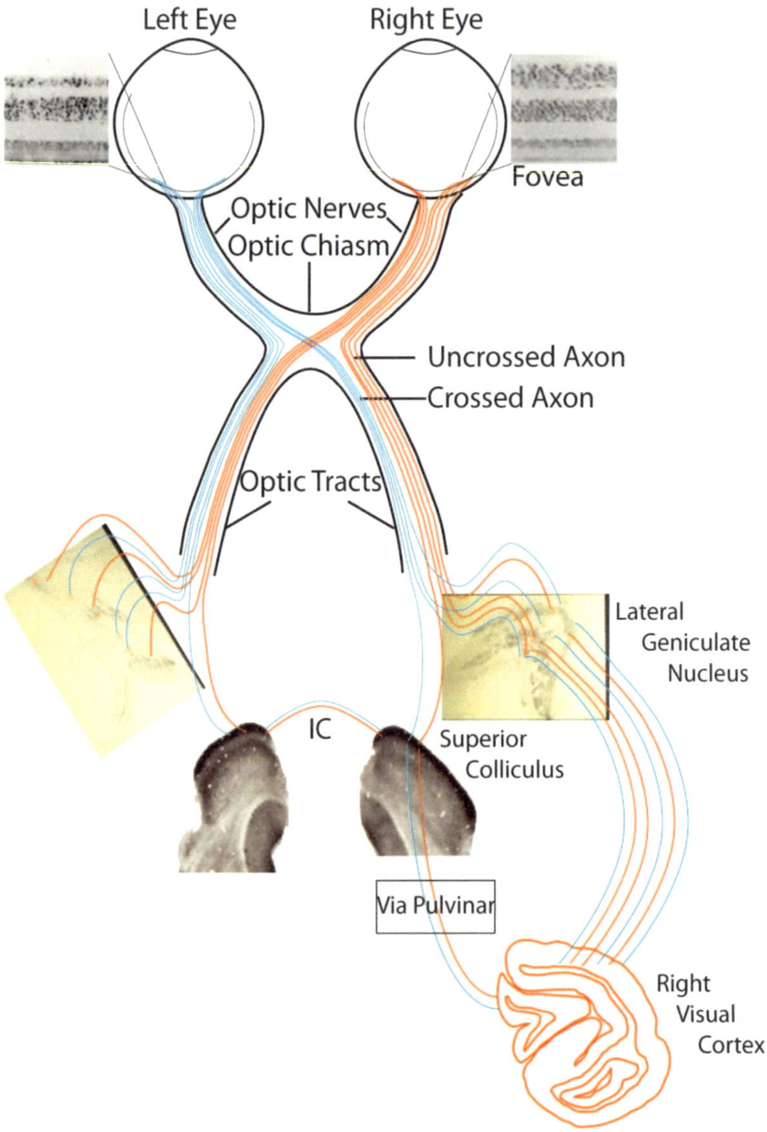

Figure 3. Hemianopia pathway: we have previously proposed an anatomical pathway for visual field recovery depicted here. Briefly, the residual left temporal retinal ganglia cells send their projections to the appropriate lateral geniculate nucleus, the function of which is not entirely clear. The retinofugal projections to the left superior colliculus remain intact with information potentially traversing to the right side through the intertectal commissure (IC), then processed through the pulvinar, and finally to the right visual cortex. The left pulvinar is severely atrophied and is unlikely to account for residual vision seen in our subjects (figure is adapted from Burke et al. [20]).

thereby supporting the predictive validity of this model. The surviving peripheral retinal ganglion cells provide the first line of residual visual capacity, but given the extent of volume and neuronal loss in the ipsilateral lateral geniculate nucleus, it is doubtful that this thalamic nucleus alone could explain vision in the blind field. The superior colliculus retains functional capacity, as revealed by cytochrome oxidase activity and relative sparing of the neuronal population [36], as well as the neuronal population in the ipsilateral substantia nigra in these monkeys. The lateral substantia nigra comprises the nigrotectal pathway, which is an important mediator of saccadic eye movements. Therefore, we have proposed that the peripheral retinal ganglion cells project to the left superior colliculus from which the information is then transferred to the right superior colliculus via the intertectal commissure to the right pulvinar and finally to the right extrastriate cortex [20, 28]. Diffusion tensor imaging (DTI) has also suggested such a retinofugal projection to the ipsilateral superior colliculus as the potential substrate for unconscious vision or blindsight in hemispherectomized patients [17, 37].

Histological data from this model, as well as that from clinical studies, suggest that residual subcortical and brain stem areas play a significant role in functional remodeling following early-life hemispherectomy. The culmination of over 20 plus-year experience with this model has shed new light on the ability of the infant brain to reorganize [20]. The application of validity criteria further shows the significant contribution to the understanding of human conditions by studying nonhuman primates. We have also applied these criteria to a nonhuman primate model of pediatric HIV infection [11]. There are relatively few pediatric simian immunodeficiency virus (SIV) models, but the ones that are available show strengths in each of the five validity criteria and provide a platform in which to test therapeutic interventions that are aimed at reducing HIV neurological dysfunction that is prevalent in the pediatric population [11, 38].

4. Conclusions

In this book, we present a series of chapters dedicated to the study of primates that range from phyletic organization, to observational and conservational efforts, to using nonhuman primates, to understand our own human condition. In their natural habitat, the interaction between humans and nonhuman primates may be contentious as monkeys may be seen as agricultural pests [1, 2, 39], something that we could only speculate Marais would argue against. Whether or not we are cognizant of his work, Marais [1] helped lay the foundation for multiple lines of research for the better understanding of our own human psyche, emphasized the need to protect and observe primates in their habitat so that we may better understand our own "soul."

Acknowledgements

This research was supported by the Behavioral Science Foundation (BSF), St. Kitts and the Natural and Engineering Research Council of Canada (NSERC RGPAS 478115-2015 to MP). We are grateful to the entire staff of BSF for excellent animal care and research assistance.

Author details

Mark W. Burke[1,2]* and Maurice Ptito[2,3,4,5]

*Address all correspondence to: mark.burke@howard.edu

1 Department of Physiology and Biophysics, Howard University, Washington, DC, USA

2 Behavioural Science Foundation, St. Kitts, Saint Kitts and Nevis

3 Department of Physiology, School of Optometry, Université de Montréal, Montréal, QC, Canada

4 Laboratory of Neuropsychiatry, University of Copenhagen, Copenhagen, Denmark

5 Department of Neurology and Neurosurgery, Montréal Neurological Institute, McGill University, Montréal, QC, Canada

References

[1] Marais E. The Soul of the Ape. New York: Human & Rousseau Publishers Ltd; 1969

[2] Palmour RM, Mulligan J, Howbert JJ, Ervin F. Of monkeys and men: Vervets and the genetics of human-like behaviors. American Journal of Human Genetics. 1997;**61**:481-488

[3] Burke MW, Palmour RM, Ervin FR, Ptito M. Neuronal reduction in frontal cortex of primates after prenatal alcohol exposure. Neuroreport. 2009;**20**:13-17

[4] Dudley R. Evolutionary origins of human alcoholism in primate frugivory. The Quarterly Review of Biology. 2000;**75**:3-15

[5] Ervin FR, Palmour RM, Young SN, Guzman-Flores C, Juarez J. Voluntary consumption of beverage alcohol by vervet monkeys: Population screening, descriptive behavior and biochemical measures. Pharmacology, Biochemistry, and Behavior. 1990;**36**:367-373

[6] Juarez J, Guzman-Flores C, Ervin FR, Palmour RM. Voluntary alcohol consumption in vervet monkeys: Individual, sex, and age differences. Pharmacology, Biochemistry, and Behavior. 1993;**46**:985-988

[7] Cloninger CR. Neurogenetic adaptive mechanisms in alcoholism. Science. 1987;**236**:410-416

[8] Burke MW, Inyatkin A, Ptito M, Ervin FR, Palmour RM. Prenatal alcohol exposure affects progenitor cell numbers in olfactory bulbs and dentate gyrus of Vervet monkeys. Brain Sciences. 2016;**6**:pii: E52

[9] Burke MW, Ptito M, Ervin FR, Palmour RM. Hippocampal neuron populations are reduced in vervet monkeys with fetal alcohol exposure. Developmental Psychobiology. 2015;**57**:470-485

[10] Papia MF, Burke MW, Zangenehpour S, Palmour RM, Ervin FR, Ptito M. Reduced soma size of the M-neurons in the lateral geniculate nucleus following foetal alcohol exposure in non-human primates. Experimental Brain Research. 2010;**205**:263-271

[11] Carryl H, Swang M, Lawrence J, Curtis K, Kamboj H, Van Rompay KK, De Paris K, Burke MW. Of mice and monkeys: Can animal models be utilized to study neurological consequences of pediatric HIV-1 infection? ACS Chemical Neuroscience. 2015;**6**:1276-1289

[12] Belzung C, Lemoine M. Criteria of validity for animal models of psychiatric disorders: Focus on anxiety disorders and depression. Biology of Mood & Anxiety Disorders. 2011;**1**:9

[13] Willner P. The validity of animal models of depression. Psychopharmacology. 1984;**83**: 1-16

[14] Wilson PJ. Cerebral hemispherectomy for infantile hemiplegia. A report of 50 cases. Brain. 1970;**93**:147-180

[15] van Empelen R, Jennekens-Schinkel A, Buskens E, Helders PJ, van Nieuwenhuizen O, Dutch Collaborative Epilepsy Surgery P. Functional consequences of hemispherectomy. Brain. 2004;**127**:2071-2079

[16] Devlin AM, Cross JH, Harkness W, Chong WK, Harding B, Vargha-Khadem F, Neville BG. Clinical outcomes of hemispherectomy for epilepsy in childhood and adolescence. Brain. 2003;**126**:556-566

[17] Ptito A, Leh SE. Neural substrates of blindsight after hemispherectomy. The Neuroscientist. 2007;**13**:506-518

[18] Chiricozzi F, Chieffo D, Battaglia D, Iuvone L, Acquafondata C, Cesarini L, Sacco A, Chiera R, Di Rocco C, Guzzetta F. Developmental plasticity after right hemispherectomy in an epileptic adolescent with early brain injury. Child's Nervous System. 2005;**21**:960-969

[19] Govindan RM, Chugani HT, Luat AF, Sood S. Presurgical prediction of motor functional loss using tractography. Pediatric Neurology. 2010;**43**:70-72

[20] Burke MW, Kupers R, Ptito M. Adaptive neuroplastic responses in early and late hemispherectomized monkeys. Neural Plasticity. 2012;**2012**:852423

[21] Clancy B, Darlington RB, Finlay BL. Translating developmental time across mammalian species. Neuroscience. 2001;**105**:7-17

[22] Petrides M, Tomaiuolo F, Yeterian EH, Pandya DN. The prefrontal cortex: Comparative architectonic organization in the human and the macaque monkey brains. Cortex. 2012;**48**:46-57

[23] Mackey S, Petrides M. Quantitative demonstration of comparable architectonic areas within the ventromedial and lateral orbital frontal cortex in the human and the macaque monkey brains. The European Journal of Neuroscience. 2010;**32**:1940-1950

[24] Fuster J. The Prefrontal Cortex. New York: Elsevier; 2008

[25] Stoerig P. Cueless blindsight. Frontiers in Human Neuroscience. 2010;**3**:74

[26] Herbin M, Boire D, Theoret H, Ptito M. Transneuronal degeneration of retinal ganglion cells in early hemispherectomized monkeys. Neuroreport. 1999;**10**:1447-1452

[27] Boire D, Theoret H, Ptito M. Stereological evaluation of neurons and glia in the monkey dorsal lateral geniculate nucleus following an early cerebral hemispherectomy. Experimental Brain Research. 2002;**142**:208-220

[28] Ptito M, Herbin M, Boire D, Ptito A. Neural bases of residual vision in hemicorticectomized monkeys. Progress in Brain Research. 1996;**112**:385-404

[29] Honda N, Matuoka T, Sawada Y, Nakano N, Suwen L, Higashimoto Y, Fukuda K, Ohgi S, Kato A. Reorganization of sensorimotor function after functional hemispherectomy studied using near-infrared spectroscopy. Pediatric Neurosurgery. 2010;**46**:313-317

[30] Bittar RG, Ptito A, Reutens DC. Somatosensory representation in patients who have undergone hemispherectomy: A functional magnetic resonance imaging study. Journal of Neurosurgery. 2000;**92**:45-51

[31] Backlund H, Morin C, Ptito A, Bushnell MC, Olausson H. Tactile functions after cerebral hemispherectomy. Neuropsychologia. 2005;**43**:332-339

[32] de Bode S, Mathern GW, Bookheimer S, Dobkin B. Locomotor training remodels fMRI sensorimotor cortical activations in children after cerebral hemispherectomy. Neurorehabilitation and Neural Repair. 2007;**21**:497-508

[33] Rutten GJ, Ramsey NF, van Rijen PC, Franssen H, van Veelen CW. Interhemispheric reorganization of motor hand function to the primary motor cortex predicted with functional magnetic resonance imaging and transcranial magnetic stimulation. Journal of Child Neurology. 2002;**17**:292-297

[34] Benecke R, Meyer BU, Freund HJ. Reorganisation of descending motor pathways in patients after hemispherectomy and severe hemispheric lesions demonstrated by magnetic brain stimulation. Experimental Brain Research. 1991;**83**:419-426

[35] Choi JT, Vining EP, Mori S, Bastian AJ. Sensorimotor function and sensorimotor tracts after hemispherectomy. Neuropsychologia. 2010;**48**:1192-1199

[36] Theoret H, Boire D, Herbin M, Ptito M. Anatomical sparing in the superior colliculus of hemispherectomized monkeys. Brain Research. 2001;**894**:274-280

[37] Leh SE, Johansen-Berg H, Ptito A. Unconscious vision: New insights into the neuronal correlate of blindsight using diffusion tractography. Brain. 2006;**129**:1822-1832

[38] Carryl H, Van Rompay KK, De Paris K, Burke MW. Hippocampal neuronal loss in infant macaques orally infected with virulent simian immunodeficiency virus (SIV). Brain Sciences. 2017;**7**:pii: E40

[39] Heymann EW, Hsia SS. Unlike fellows—A review of primate-non-primate associations. Biological Reviews of the Cambridge Philosophical Society. 2015;**90**:142-156

Chapter 2

An Overview of the Primates

Colin Groves

Additional information is available at the end of the chapter

http://dx.doi.org/10.5772/intechopen.70264

Abstract

The order Primates, to which humans belong, is one of the best-known mammalian orders, but there is still much to be learned about its phylogeny and taxonomy. It is clear by now that there are two suborders, Strepsirrhini and Haplorrhini, but beyond that there is still a lot of controversy and misunderstanding including how to operationalise the evolutionary species. The example of the Old World Monkey tribe Cercopithecini is treated in some detail.

Keywords: primates, taxonomy, phylogeny

1. Introduction

The order Primates, constituting one of about 20 orders of placental mammals, is most closely related to the orders Scandentia (treeshrews) and Dermoptera (colugos). The order consists of lemurs, lorises, bushbabies, tarsiers, New World monkeys, Old World monkeys, apes and humans.

It is hard to diagnose the order Primates, because evolution is an ongoing process, and those features which characterised the ancestors of any group of organisms may well have changed in their descendants; as the order Primates originated some 65 million years ago, there has been ample time for change. This was pointed out in detail by Szalay et al. [1], who suggested a number of conditions that probably constituted the Primate morphotype (i.e., the ancestral state of primates); if we select just those features which are diagnostic of modern primates, we can list the following:

1. Auditory bulla formed by the petrosal (in other mammals with an ossified auditory bulla, the bulla is formed from the alisphenoid, or from the tympanic, or from one or more of a group of special bones, the entotympanic complex).

2. The underside of the ends of the digits has a specialised nerve ending, Meissner's corpuscles (this is convergent on some of the arboreal marsupials).
3. Hallux is divergent from other toes, and is supplied with a flat nail instead of claws (this, obviously, is modified in humans). This results in a grasping foot, and is, again, convergent on some arboreal marsupials. In some primates, the hand has specialised grasping abilities as well, and in most but not all primates all the digits, not just the hallux, have flat nails.
4. Penis hangs free from the body wall (this is convergent on the order Chiroptera, the bats; in other mammals, it is enclosed within a prepuce).

We, when thinking about primates, automatically tend to think of convergent and frontated orbits, and of binocular vision, but these are in fact fairly widespread among other mammals, both in placentals and marsupials.

It is much easier to define the subgroups of Primates, as follows:

1.1. Suborder Strepsirrhini

They are characterised by the possession of a rhinarium (moist nasal tip, continuous, via a split upper lip, with the gums); by having a tapetum lucidum (reflective layer) behind the retina in the eye; and by having an epitheliochorial placenta (maternal and foetal vascular systems well separated, and the embryo/foetus is nourished largely by uterine glands). They are the lemurs, lorises and galagos (bushbabies).

1.2. Suborder Haplorrhini

They lack a rhinarium; they lack a tapetum, but have a fovea and macula in the retina; they have a haemochorial placenta (embryonic/foetal tissue invades the maternal bloodstream). They are the tarsiers, monkeys, apes and humans.

It is not easy to say which of these diagnostic features may be derived, and which of them may be primitive. Almost certainly, the possession of a rhinarium is a primitive retention, but there are arguments for each of the others being evolutionarily derived features.

The two suborders separated well back in the Cenozoic, probably in the Palaeocene, although some estimates place their split even as far back as the Middle Cretaceous (see [2]). The point to emphasise is that they are clades—that is to say, each is monophyletic. Monophyly (descent from an exclusive common ancestor) is the essential basis for systematics above the species level. Some older Primate classifications, widely adopted up until the 1980s and still sometimes seen in textbooks even as late as the early 2000's, were recognised, as the two suborders, Prosimii (strepsirrhines plus tarsiers) and Anthropoidea (monkeys, apes and humans); but this tells us nothing about relationships, which is what systematics is all about, as the group called Prosimii is defined by nothing except retention of some primitive characters (mostly, relatively small brains).

2. The Strepsirrhini

The Strepsirrhini are the lemurs, lorises and galagos. They divide into two monophyletic clades, which, conveniently, are geographically separated: one in Madagascar, one in mainland Africa and tropical Asia. These separated from each other in the Palaeocene or Eocene (the TTOL calculates a median time of separation at 58.5MA). Many classifications place the Malagasy lemurs into one infraorder, Lemuriformes, and the non-Malagasy lorises and galagos into another, Lorisiformes; but the evidence indicates that the Malagasy group divided again very early on, one of the clades containing just a single living species, the aye-aye, *Daubentonia madagascariensis*, the other containing all the other Madagascar lemurs. The TTOL gives a median separation time of 54 MA, which is very close to the time of separation of the Malagasy group as a whole from the Lorisiformes, and so much earlier than the other groups of Malagasy lemurs started to diversify (see below) that the aye-aye is now generally removed from the Lemuriformes and placed in a third infraorder, Chiromyiformes. Thus, we split the Strepsirrhini into three infraorders: Lorisiformes, Lemuriformes and Chiromyiformes.

There is, in fact, some suggestion that *Daubentonia* may be the sister-group of the Late Eocene genus *Plesiopithecus*, which would imply that the Lemuriformes and Chiromyiformes dispersed separately to Madagascar from the African mainland [3]. This would actually make a great deal of sense, because the Lemuriformes started to separate into families in Madagascar around the Middle-Late Oligocene, at the same time as the families of the other endemic Malagasy mammals (euplerid carnivores, tenrecs and nesomyine rodents), which might suggest that they had all arrived together in a new, mammal-free island and began to diversify into vacant niches.

The families of Lemuriformes are Lemuridae, Cheirogaleidae, Indriidae and Lepilemuridae (**Figure 1**).

Figure 1. Giant Bamboo Lemur, *Prolemur simus*.

Figure 2. Bengal Slow Loris, *Nycticebus bengalensis*.

There is only one family in the Chiromyiformes: Daubentoniidae. Within the Lorisiformes, division into two families, Lorisidae and Galagidae, is usual; but the two African genera ascribed to Lorisidae (*Perodicticus* and *Arctocebus*), stubbornly refuse to align with the two Asian genera (*Loris* and *Nycticebus*) (**Figure 2**) in most molecular studies [3], and there is probably a case for recognising them as a separate family, Perodicticidae, implying that they acquired their 'slow climbing' adaptations quite independently of each other.

3. The Haplorrhini

The Haplorrhini, like the Strepsirrhini, divide into two clades, classified as infraorders: Tarsiiformes and Simiiformes. The latter infraorder is more usually called Anthropoidea, a term which has in the past been used confusingly, so is better avoided (whereas, there are rules for the nomenclature of species, genera and families, there are none for taxonomic ranks above the family-group level).

It was the Tarsiiformes which caused all the angst over the Primate suborder question in the latter half of the twentieth century. From the point of view of that most solipsistic of primates, *Homo sapiens*, the Tarsiiformes are irredeemably primitive because they have small brains—hence, so the argument went, they should be bundled into the Prosimii along with those other small brained primates, the lemurs, lorises and galagos. A taxonomic philosophy that emphasises relatedness, however, puts them into the Haplorrhini, because there is absolutely no doubt that they share derived conditions with monkeys and apes. Molecular phylogenies amply confirm that this is where they truly belong [4, 5].

There is only a single family, Tarsiidae, of Tarsiiformes, but this divides into three genera [6]. So far, only the genus *Tarsius* (the Eastern Tarsiers, from Sulawesi and offshore islands) has been studied taxonomically, and there turn out to be quite a large number of species; the other two genera, *Cephalopachus* (Western Tarsiers, from Borneo, Sumatra and intermediate islands) and *Carlito* (Philippine Tarsiers), are assigned a single species each, but this is only because there have been no detailed taxonomic surveys of them (**Figure 3**).

The Simiiformes again divide into two clades, Platyrrhini and Catarrhini, alternatively classified as parvorders or, sometimes, left unranked (there is much dissension about unranked

Figure 3. Maros tarsier, *Tarsius fuscus*.

taxonomic systems; in the opinion of most taxonomists, ranks are useful, especially if tied to some time-depth scheme). The Catarrhini are distinguished in particular by the reduction of the premolars in both jaws from three to two, and the 'deflation' of the auditory bulla, with the tympanic ring extended into a tube. The Platyrrhini retain the primitive three premolars, inflated bulla and ring-like tympanic, and it has proved difficult to find any derived character states which they share, although molecular research confirms their monophyly; most recently, however, it has been shown that contact between parietal and zygomatic (malar) bones is a truly derived platyrrhine condition [7].

The Platyrrhini are the New World monkeys; all of them occur in South America, some of them extending north into Central America, even as far as southern Mexico. There are three families of Platyrrhini: Cebidae (including the marmosets and tamarins, at one time assigned to their own family), Atelidae and Pitheciidae (**Figures 4–6**).

The Catarrhini all live in the Old World. There are three families of Catarrhini: Cercopithecidae, belonging to the superfamily Cercopithecoidea (the Old World monkeys) (**Figure 7**), and the Hylobatidae (gibbons, of Southeast Asia) (**Figure 8**) and Hominidae (great apes and humans) (**Figures 9** and **10**), these latter two belonging to the superfamily Hominoidea. Molecular calculations of the separation time of the two families of Hominoidea tend to fall in the late Early Miocene, well below the criterion laid down by Goodman et al. [8] and Groves [9], that families should have separated around the Oligocene-Miocene boundary, and it is probable that the two hominoid families should be reduced to a subfamily rank within a single family, as indeed proposed by Goodman et al. [8].

Figure 4. Colombian night monkey, *Aotus griseimembra*, a member of the Cebidae.

Figure 5. Pied Tamarin, *Saguinus bicolor*, a member of the Cebidae.

Figure 6. Northern red howler monkey, *Alouatta seniculus*, a member of the Atelidae.

Figure 7. A female olive baboon, *Papio anubis*, from Kenya (a member of the Cercopithecidae), filling her cheek pouches with food.

Figure 8. Kloss gibbon, *Hylobates klossii*.

http://dx.doi.org/10.5772/intechopen.70264

Figure 9. Sumatran orangutan, *Pongo abeli*.

Figure 10. Western Gorilla, *Gorilla gorilla*.

4. The species of primates

There are many more species of lemurs that had previously been appreciated [10], and the list is still growing. Similarly, the number of species in some platyrrhine genera is growing (see, for example, [11]), and there are many more species of Asian primates than had previously been recognised [12]. This is much more that simply recognising supposedly 'trivial' differences as diagnostic of species: where we have knowledge of them, all the species have ecological and/or physiological significance. This brings us to ask the important question: what actually is a species?

There has been a lot of discussion about 'the species question' over the past 20–30 years, and several surveys have converged on the essence of what we mean by species: they are evolutionary lineages [9, 13, 14]. Species thus have a real existence. This settles the ontological status of the species concept, but it does not necessarily solve the question of how to recognise them; the most logical way of defining species operationally is by the so-called Phylogenetic Species Concept: 'A species is the smallest population or aggregation of populations which has fixed heritable differences from other such populations or aggregations' [15]. This definition has three strands:

1. Species are populations (or aggregations of populations). They are not, for example, segments of populations. This can generally be observed in the field, though in the museum or on the laboratory bench it must be inferred.
2. The differences are heritable. They may thus be differences observable in the genome; or they may be differences in morphological characters or in behavioural characters, in which case a heritable (genetic) basis is only inferential, although such a basis tends to be very strongly implied.
3. The differences are fixed. This is a geneticists' term meaning that one allele, or even just one base-pair, occurs in 100% of one population, but in 0% of another. That is to say, the two populations are absolutely (diagnosably) different.

Many people have found difficulty coming to terms with the evolutionary species itself, let alone with the phylogenetic/diagnosability species criterion, feeling that a species is not a 'real' species unless, for example, it does not or cannot interbreed with other species (the notion which some of us were brought up with, and which unfortunately is still taught in many schools, even in university courses). The work of Christian Roos and his colleagues has shown that, in fact, there has been very widespread interbreeding and gene exchange between different species during their evolution (see, for example, [16]).

5. The case of the Cercopithecini

The strepsirrhines, the tarsiers and the platyrrhines have been subjected to critical taxonomic revision over the past 15 years or so. What of the catarrhines?

The Asian catarrhines (gibbons, the orangutan, colobines and the genus *Macaca*) have in most cases been carefully examined taxonomically, and work on them continues [12]. The macaques and langurs, especially, both need further attention. New species of macaques have been described, and the question of the gene flow between the long-tailed macaque *Macaca fascicularis* and the rhesus macaque *Macaca mulatta*, where their ranges meet (in mainland Southeast Asia) is an ongoing focus of research.

This leaves the African catarrhines, particularly the Old World Monkeys (superfamily Cercopithecoidea, family Cercopithecidae). The Cercopithecidae are divided into two subfamilies. The African members of the subfamily Colobinae (the leaf eating monkeys) need further attention, both genetically and morphologically. Within the other subfamily, Cercopithecinae (omnivorous monkeys with cheek pouches), the taxonomy of the baboons (*Papio*), of the tribe Papionini, is receiving continual attention and has already yielded unexpected insights [17]. The related genera *Cercocebus*, *Lophocebus*, *Rungwecebus*, *Theropithecus* and *Mandrillus* (mangabeys, geladas and mandrills) remain to be investigated in depth. But, the outstanding case of a group that requires attention is the genus *Cercopithecus* and its relatives, which together form the tribe Cercopithecini.

Traditionally, the Cercopithecini have been divided into four genera, *Cercopithecus* (the rain forest living guenons and their relatives), *Erythrocebus* (the patas monkey), *Miopithecus* (the talapoin monkey) and *Allenopithecus* (the swamp monkey); while over the past quarter century, the further separation of *Chlorocebus* (savanna monkeys, the vervet group) and *Allochrocebus* (terrestrial forest monkeys: the former *Cercopithecus lhoesti*, *preussi* and *solatus*) has been increasingly recognised. This six-genus scheme is adopted in the latest compendium [18], and we may summarise them as follows:

1. *Cercopithecus*. These are colourful, mostly rainforest, monkeys. The *C. nictitans/mitis* group, the *C. mona/campbelli/pogonias/wolfi* group and the *C. petaurista/cephus/ascanius* group are widespread from West Africa through central Africa into at least Uganda, with some 'subspecies' of the *C. mitis* group extending into Ethiopia, and even South Africa (**Figures 11–13**). *C. diana* and its sister species *C. roloway* are confined to West Africa. *C. neglectus* is found in central Africa and extends into Kenya and Ethiopia. *C. hamlyni* and its sister species *C. lomamiensis* are confined to the east of the Democratic Republic of Congo. The enigmatic *C. dryas* is found in small areas in the DRC.

2. *Allochrocebus*. These are terrestrial rain forest monkeys. There are three species: *A. lhoesti* from the eastern DRC, *A. preussi* from Cameroon and *A. solatus* from south-eastern Gabon.

3. *Chlorocebus*. Several Savanna woodland species, including *C. sabaeus* (the Green monkey of West Africa), *C. aethiops* (the Grivet monkey of Ethiopia) and *C. pygerythrus* (the Vervet monkey of eastern and southern Africa).

4. *Erythrocebus*. The Patas monkey of the grasslands of mainly West Africa, all belonging to a single species, *E. patas*.

5. *Miopithecus*. Two very small species known as Talapoin monkeys, from West-central Africa (Cameroon, upon, Congo Republic and Angola). Unlike the previous genera, talapoins

have periodic sexual swellings in females, a link to the Papionini which also have sexual swellings.

6. *Allenopithecus*. The Swamp monkey of the lower Congo River; this also has periodic sexual swellings, and in other morphological characters it recalls the Papionini.

Until just a few years ago, everybody seems to have settled down with this six-genus scheme, a few authors suggesting particular links between some of the genera. Groves [19] noted cranial characters shared between *Chlorocebus* and *Erythrocebus*, suggesting that these two genera form a clade: their similarities could perhaps relate to their general non-forest adaptations, which might suggest convergence as much as sister-group status. Within *Cercopithecus*, sister-group relationship between the *C. mona* and *C. cephus* groups and between *C. hamlyni* and the *C. lhoesti* group (the latter now recognised as the genus *Allochrocebus*) was postulated by Groves [19], who also drew special attention to the mystery species *C. dryas* and *C. salongo*, although it was not recognised at the time that they are in fact, respectively, the juvenile and adult of one and the same species (interestingly, the skull of the juvenile type specimen of *C. dryas* has vervet-like resemblances, whereas the available adult skulls seem not to). Further complications in the Cercopithecini were noted by Groves [9].

Figure 11. Putty nosed monkey, *Cercopithecus nictitans*, a member of the *C. mitis* group.

Figure 12. Wolf's Mona, *Cercopithecus wolfi*.

Figure 13. Red-tailed monkey, *Cercopithecus ascanius*.

Most recently, a remarkable study by Guschanski et al. [20] used museum samples to collect mtDNA, and incorporated carefully chosen samples from GenBank, including nDNA as well as mtDNA. This has clarified some relationships, but at the same time added quite unexpected complexity to others.

In the phylogeny of Guschanski et al. [20], there are four major 'speciation event' time periods. The initial split within the Cercopithecini is between *Miopithecus* and the rest, and a second split between *Allenopithecus* and the rest comes very quickly afterwards; both are around 9.5 Ma, and this near-trifurcation constitutes the first speciation event. That these two genera are the most divergent of the Cercopithecini is certainly not unexpected, although the order of branching perhaps is.

The second speciation event, between about 6.5 and 7.5 Ma, sees four branches separate, and these separations are most unexpected:

1. *Chlorocebus* plus *Cercopithecus hamlyni*. Within the *Chlorocebus* cluster, in the third-speciation event, three lineages diverge: *Allochrocebus solatus*, *C. sabaeus*, and a clade containing not only the other species of *Chlorocebus*, but also *Cercopithecus dryas*. These relationships were totally unsuspected, and raise several questions: is *C. hamlyni* not a member of the genus *Cercopithecus* after all? Is the genus *Allochrocebus* non-monophyletic? Is *C. dryas* likewise not a member of the genus *Cercopithecus*, but is it instead a 'vervet' that has recolonised a rainforest niche, and has it somehow undergone hypermorphosis, in that the juvenile skull reveals its affinities, whereas adult skulls have changed?
2. A clade consisting of the *C. mona*, *C. neglectus* and *C. diana* groups; in this, *C. diana* and *C. roloway* do not cluster together, but are the first and second to separate within the clade. Is the *C. diana* group, therefore, non-monophyletic?
3. *Erythrocebus*. Patas monkeys are uncontroversial.
4. A *C. cephus* plus *Allochrocebus preussi*/*lhoesti* subclade, and the *C. nictitans*/*mitis* group as the other. This raises further questions about the phylogenetic position and the monophyly of the genus *Allochrocebus*.

The third-speciation event sees the major branchings within these groups. There is nothing surprising about these (given the unexpected composition of some of the major clades), but we can notice that, within clade 2, *C. mona* and *C. campbelli* separate strongly (at or before the third-speciation event) from the *C. pogonias* cluster; and, within clade 4, the third-speciation event splits are *C. mitis opisthostictus*, *C. mitis mitis* plus *C. nictitans*, and the rest of *C. mitis*, i.e., the presumed species *C. mitis* is non-monophyletic.

But, remarkably, there is some noteworthy discrepancy between the new mtDNA data and the nDNA. Most notably, *C. hamlyni* associates with *Chlorocebus* in mtDNA but is part of *Cercopithecus*, although the most distinct branch, in nDNA; and it is the *A. lhoesti* group which is associated with *Chlorocebus* in nDNA. Again, *Erythrocebus* is associated with *Chlorocebus* in nDNA but is nested within *Cercopithecus* in mtDNA. These discrepancies are noted by the authors who suggest periods of hybridisation with nuclear swamping as a possible explanation.

http://dx.doi.org/10.5772/intechopen.70264

How would this work? An example might be as follows. A population of very early *Chlorocebus* (proto-vervets) was invaded by males from a population of proto-*Cercopithecus*. The invading males were dominant over the indigenous males, and so mated with all the indigenous (proto-vervet) females. The male hybrids were again dominated by the proto-*Cercopithecus* males, which therefore were able to mate with the female hybrids, generating backcrosses which were 75:25 in nuclear DNA; in the next generation, it happened again. In the end, we have a population which in essence is proto-*Cercopithecus*, but retains the (matrilineally inherited) mtDNA of *Chlorocebus*, which, over the course of time, developed its own characteristic features and became *Cercopithecus hamlyni* (presumably also *C. lomamiensis*), and so on.

And then there is the taxonomy. If we accept the Miocene/Pliocene boundary as the cut-off point for genera, then (apart from the uncontroversial *Erythrocebus*, *Miopithecus* and *Allenopithecus*) there are, according to the mtDNA scenario, at least five genera involved: *Chlorocebus* (including *C. solatus* and *C. dryas*), one for *C. hamlyni*, one for clade 2, and one each for the two subclades of clade 4. In addition, the relationships within clade 2 are so unexpected that more genera might be involved. But the discrepancies noted by the authors, between nDNA and mtDNA (noted above), dictates caution before we start to revise the taxonomy wholesale.

6. Where to now?

The old taxonomic certainties with which we have been comfortable for so many years have been shaken to the core. Species that do not interbreed are somewhat of a rarity. Taxonomy, at least above the species level, is based severely on monophyly. Inter-relationships between families, between genera, and between species are often not at all as we thought they were.

Taxonomy is a dynamic science. There is no excuse for a conservative attitude: disagreeable as it may seem, we have to constantly be prepared for new revelations, overturning the old certainties.

Author details

Colin Groves

Address all correspondence to: colin.groves@anu.edu.au

Australian National University, Canberra, Australia

References

[1] Szalay FS, Rosenberger AL, Dagosto M. Diagnosis and differentiation of the order primates. Yearbook of Physical Anthropology. 1987;**30**:75-105

[2] TTOL (TimeTree of Life). 2016. http://www.timetree.org/

[3] Chatterjee HJ, Ho SYW, Barnes I, Groves C. Estimating the phylogeny and divergence times of primates using a supermatrix approach. BMC Biology. 2009;**9**:259. DOI: 10.1186/1471-2148-9-259

[4] Hartig G, Churakov G, Warren WC, Brosius J, Makałowski W, Schmitz J. Retrophylogenomics place tarsiers on the evolutionary branch of anthropoids. Scientific Reports. 2013;**3**:1756. DOI: 10.1038/srep01756

[5] Ziętkiewicz E, Richer C, Labuda D. Phylogenetic affinities of tarsier in the context of primate Alu repeats. Molecular Phylogenetics and Evolution. 1999;**11**:77-83

[6] Groves C, Shekelle M. The genera and species of *Tarsiidae*. International Journal of Primatology. 2010;**31**:1071-1082

[7] Fulwood EL, Boyer DM, Kay RF. Stem members of Platyrrhini are distinct from catarrhines in at least one derived cranial feature. Journal of Human Evolution. 2016;**100**:16-24

[8] Goodman M, Porter CA, Czelusniak J, Page SL, Schneider H, Shoshani J, Gunnell G, Groves CP. Toward a phylogenetic classification of primates based on DNA evidence complemented by fossil evidence. Molecular Phylogenetics and Evolution. 1998;**9**:585-598

[9] Groves C. Primate Taxonomy. Washington: Smithsonian Institution Press; 2001

[10] Mittermeier RA, Louis EE, Richardson M, Schwitzer C, et al. Lemurs of Madagascar. Illustrated by Nash SD. 3rd ed. Conservation International; 2010 ISBN: 978-1-934151-23-5

[11] Marsh LK. A taxonomic revision of the Saki monkeys, *Pithecia* Desmarest, 1804. Neotropical Primates. 2014;**21**(1):1-163

[12] Roos C, Boonratana R, Supriatna J, Fellowes JR, Groves CP, Nash SD, Rylands AB, Mittermeier RA. An updated taxonomy and conservation status review of Asian primates. Asian Primates Journal. 2014;**4**(1):2-38

[13] De Queiroz K. Species concepts and species delimitation. Systematic Biology. 2007;**56**:879-886

[14] Mayden RL. A hierarchy of species concepts: The denoument in the saga of the species problem. In: Claridge MF, Dawah HA, Wilson MR, editors. Species: The Units of Diversity. London: Chapman and Hall; 1997. pp. 381-423

[15] Groves C, Grubb P. Ungulate Taxonomy. Baltimore: The Johns Hopkins University Press; 2011

[16] Wang B, Zhou X, Shi F, Liu Z, Roos C, Garber PA, Li M, Pan H. Full-length numt analysis provides evidence for hybridization between the Asian colobine genera *Trachypithecus* and *Semnopithecus*. American Journal of Primatology. 2015;**77**:901-910

[17] Zinner D, Wertheimer J, Liedigk R, Groeneveld LF, Roos C. Baboon phylogeny as inferred from complete mitochondrial genomes. American Journal of Physical Anthropology. 2013;**150**:133-140

http://dx.doi.org/10.5772/intechopen.70264

[18] Butynski TM, Kingdon J, Kalina J. Mammals of Africa, II: Primates. London: Bloomsbury Publishing; 2013

[19] Groves CP. A Theory of Human and Primate Evolution. Oxford: Clarendon Press; 1989

[20] Guschanski K, Krause J, Sawyer S, et al. Next-generation museomics disentangles one of the largest Primate radiations. Systematic Biology. 2013;**62**:539-554

Chapter 3

Preliminary Observations of Infant Ontogeny in the Philippine Tarsier (*Tarsius syrichta*) and the First Description of Play Behaviour and Its Ontogeny in Tarsiers

Milada Řeháková

Additional information is available at the end of the chapter

http://dx.doi.org/10.5772/intechopen.70512

Abstract

In this paper, I present the first observations on the ontogeny of tarsier infants during their first and second months and of the mother-offspring bond from birth until separation. Tarsier mother-infant pairs were observed to be solitary. Maternal care involved a cache and carry strategy with the infant spending around 66% parked alone whilst the mother was foraging. I observed the following behaviour: rest, grooming, suckling, play, being carried, jumping and climbing. Percentage of time spent in each activity was calculated. Furthermore, play was observed to include locomotor-, social- and object-directed behaviours. Acoustic communication between mother and infant was very frequent containing four different signals. During the first days after birth, the mother transported the infants about, but within a week, the infants started climbing branches on their own. For the first three weeks, the mothers carried infants orally after which they started to follow the mother over shorter distances, although for longer distances, up to 50 days of age, the mothers continued to carry them. The inter-birth interval was around 1 year, but the infants from the previous birth continued to sleep with the mother for up to 10 months.

Keywords: Philippine tarsier, *Tarsius syrichta*, ontogeny, behaviour, maternal care, play

1. Introduction

Tarsiers are small nocturnal primates from Southeast Asia, restricted to a few islands in the Sunda Strait (off Indonesia, Malaysia and Brunei) and on some islands of the Philippines. Only one of the extant species, Philippine tarsier (*Tarsius syrichta*), lives in the Philippines.

Despite a recent increase in studies of tarsiers, they remain the least studied primates. This is unfortunate given the pivotal phylogenetic position of tarsiers between strepsirrhines and other haplorrhines [1].

Sociality, mating system and group composition vary among tarsier species. Some species cluster together in small groups. For example, spectral tarsiers (*Tarsius spectrum*) exhibit substantial amount of gregariousness and sleep together in small family groups [2–4]. Similarly, Dian's tarsier (*Tarsius dianae*) has been observed sleeping in small family groups involving either monogamously or polygynously bonded adults [5]. Finally, Lariang tarsiers (*Tarsius lariang*) live in small groups that include multiple individuals sleeping together [6, 7]. In contrast, western tarsier (*Tarsius bancanus*) [8] and Philippine tarsier [9–11] sleep and forage alone.

Apart from relatively well-studied spectral tarsier, little is known about the behaviour of tarsiers. The small size and cryptic nature of tarsiers make field observations, especially on the development of behaviour, difficult to obtain. In spectral tarsiers, birth is given to single offspring seasonally, mostly in April–May [12]. Although the data are more limited, a similar pattern appears to be present in the Philippine tarsier with births occurring between April and July [11]. At birth infants can weigh up to one-third of the body weight of mothers [13]. Being strictly insectivorous (or carnivorous) tarsier mother cannot transport the infant whilst hunting. Instead of being continually transported, infants are parked on a branch (or in a nest in the case of spectral tarsier) and left alone whilst the mother is foraging [2, 3]. The mothers return to the infants at regular intervals to feed them. In addition, on some visits the mothers may transfer them to another parking site. Variations on this kind of parking strategy also occur in several nocturnal strepsirrhines, such as *Galago*, *Microcebus*, *Cheirogaleus*, *Varecia*, *Lepilemur*, *Otolemur*, *Phaner* or *Arctocebus*, *Nycticebus* and *Perodicticus* (reviewed in [14]).

So far, study of infant and maternal behaviour in free-ranging tarsiers has been limited to the spectral tarsier [3, 4, 12]. In this species, the infant is parked and left alone for 43% of the time during the first 3 months. However, spectral tarsiers are a social species living in family groups, with alloparental care (i.e. allocare) provided by other group members, especially subadult females, for some of the time when the mother was absent. In contrast, for the solitary Philippine tarsier, preliminary data indicate that parental care is mainly provided by mother [11]. Little else is known about maternal behaviour and the development of behaviour in the young. For example, play behaviour is an important component of development in nearly all primate, and many non-primate, species [15, 16]. Yet, apart from a short note on play behaviour, present as a part of allocare in spectral tarsiers [2, 12] and in captive western tarsiers [17], play behaviour has not been described in detail in tarsiers.

In this paper, I present the preliminary observations on the ontogeny of Philippine tarsier infants during the first and second months of the age, as well as on the mother-offspring bond from birth until separation. Furthermore, I provide the first description of play behaviour in tarsiers.

2. Methods

2.1. Study site and animals

The observations on the Philippine tarsier (*T. syrichta fraterculus*) were conducted on Bohol Island, Philippines, as a part of a broader study. The landscape occupied by the tarsiers was composed of karst, narrow ravines and steep stony slopes with numerous cracked rocks covered by secondary forest. Two mother-infant pairs were observed during April and June 2009 at the Philippine Tarsier Sanctuary in Corella, Bohol, Philippines (9°41′ N, 123°57′ E) (elevation 100–200 m). The first infant (infant I) was born to mother I in a semi-captive enclosure [18] on 17 April 2009 and observed until 6 June 2009. The second infant (infant II) was born to a free-ranging radio-collared mother (mother II) on 30 April and was observed until 4 June 2009 when it was predated on by an unknown predator [19]. Additionally more limited observations were made on several other mother-infant pairs in Bilar, Bohol (9°44′ N, 124°06′ E) (elevation 220–380 m). One mother-infant pair (infant III and mother III) was observed on 22 May 2010, after they were rescued from poachers. During afternoon, they were kept in a cage, and in the evening, they were released in Forest Academic Research Area, limiting my observations to just the day when they were in the cage. Five additional infants (IV, V, VI, VII and VIII) were observed to be born to the same mother in a forested area at the Habitat Bohol Conservation Centre. Infant IV was first spotted on 22 May 2013. As it was being transported by the mother, I estimated its age to be around three weeks (based on behaviour; see below). They were irregularly checked until 5 March 2014. On 2 April 2014, the mother was found sleeping alone, suggesting that they had separated. Infant V was first spotted on 3 May 2014 and although very small was able to climb on its own a little on a branch; I estimated its age to be about 1 week. This infant was checked until 21 January 2015. Given that the mother was found alone on 20 March 2015 on the sleeping site, again it seems likely they had separated by this date. Infant VI was first spotted on 11 May 2015 and was still with the mother until 11 November 2015. The mother was spotted alone on 16 January 2016. Infant VII was first spotted on 26 April 2016 and was still with until 24 February 2017. Unfortunately, due to the lack of observers available in the field, period leading up to separation in those two infants was not sampled regularly. On the 11th of May the female was spotted with a newborn infant VIII and observed only for four days due to lack ofobservers and publication deadline. Observation periods are summarized in **Table 1**.

2.2. Data collection and analyses

We started observing infants I and II in the evening of the day when they were born. Both infants were observed irregularly during the study period by two observers. The mother-infant pairs were observed during evenings (starting between 16:45 and 18:00), from when they woke up and then followed until the infant was parked. Thereafter, the infant was observed on the parking spot until the mother returned, and then, if possible, the pair was followed again. Due to the difficult terrain and dense vegetation, it was not possible to locate the infant on every parking spot either when first deposited or during the night. Due to

Infant	Start of observation	End of observation/still sleeping with mother	Mother-infant separated
I	14th April 2009	6th June 2009	Unknown
II	30th April 2009	4th June 2009	Predated
III	22nd May 2010	22nd May 2010	Unknown
IV	22nd May 2013	5th March 2014	2nd April 2014
V	3rd May 2014	21st January 2015	20th March 2015
VI	11th May 2015	11th November 2015	16th January 2016
VII	26th April 2016	24th January 2017	Unknown
VIII	11th May 2017	14th May 2017	

Table 1. Observation periods of eight infants.

the difficulties in relocating infants later in the night, the results obtained during the study were biased towards the evening and, therefore, have to be considered as preliminary. The observed behaviour was recorded continuously using voice recorder. Infant carrying and play behaviour were photographed and video-recorded using a night vision video camera (SONY HDR-SR11). Vocalizations were recorded using digital recorders with built-in microphones (Olympus LS-10 and LS-11). In total we obtained 8.9 h of observation over 5 days for infant I and 65.1 h over 23 days for infant II. The length of focal follows was between 13 min and 12.3 h (the mean was 2 h 38 min). Infant III was observed during daytime from 10:30 and then for 1 h (18:30–19:30) in the evening after being released into the wild. The vocalizations from this infant were analysed. Mother-infant pairs IV, V, VI, VII and VIII were observed irregularly during the daytime and evening. In these pairs we focused mainly on recording the period when they were still sleeping, during which they were in contact or in close proximity, patterns of association not previously known for this species.

We recorded behaviour of the infants during individual focal follows [20] and developed a basic ethogram (see Section 3). We also recorded the distance between mother and infant: contact (physical contact), proximity (<3 m) and alone (>3 m). For further analyses the data were pooled together for infants I and II. The relative distances, time period and percentage time spent in each behaviour were calculated.

3. Results

The day before the mother I gave birth, she stayed very high (above 5 m) on the tree in the semi-captive enclosure. In contrast, the free-ranging mother II stayed no more than 1 m above the ground. Indeed, a week before giving birth, she was observed on occasion sleeping on the ground.

The mother-infant pairs were observed to be solitary. On two occasions (one each for each pair) another individual (most likely a male given its size) came within about 5 m of the mother-infant pair. The intruding individual was immediately chased away by the mother, who was vigorously jumping towards the intruder and emitting several loud calls. Another similar encounter was observed for mother-offspring pair V. In this case, the intruding tarsier was observed at the top of the tree, more than 5 m from the mother-infant pair. It could not be ascertained whether the intruder was an unrelated adult male or the older offspring from the previous year who remained in the area.

3.1. Infant behaviour and time budgets

Tarsiers woke up shortly after our arrival or were already awake. After waking up the mother and infant spent some time on the sleeping site before they moved to another spot—where the infant was parked. During our observations the following behaviours were recorded. The activities and behaviours listed below are mutually exclusive:

Resting: an infant is awake but unmoving whilst sitting on a branch (**Figure 1**), a stance that was evident either shortly after waking up in the evening or during the night.

Grooming: an infant being groomed by the mother (only on two occasions was self-grooming by the infant recorded).

Suckling: an infant being nursed by the mother (**Figure 2**).

Play: an infant was considered playing when the movement or the body posture was awkward, exaggerated or incomplete; or the speed, aiming or accuracy of the movements was relaxed (criteria consistent with the definition of play developed by Burghardt [21]). See below for more details (Video 1).

Figure 1. A tarsier infant (infant II) resting on a branch in contact with its mother (photo by M. Řeháková).

Figure 2. A tarsier infant (infant VIII) suckling (photo by M. Řeháková).

Carried: an infant being carried by the mother and transported from one location to another (**Figure 3**).

Jumping: an infant jumping from one branch to another.

Climbing: an infant climbing up or down a branch.

Furthermore, we reported if the infant was:

Figure 3. (a–c) A tarsier infant (infant I) being transported in its mother's mouth and holding slightly by its tail around the mother's neck (photo by P. Slavík).

With mother: mother is in contact or in proximity of the infant.

Parked: an infant is left on a parking spot whilst the mother goes for foraging. Also, when parked if the infant was visible, we recorded any of the above-listed behaviours that could occur.

And we recorded if the infant was:

In contact: in physical contact with the mother.

In proximity: within 3 m to the mother. Three metres was a maximal distance at which I could reliably identify an individual in the dense vegetation. Moreover, at this distance it is likely that the infant and mother were either in visual or acoustic contact.

Alone: more than 3 m from mother, only when the infant was parked.

Based on 74 h during which infants I and II were observed, infants spent 66% of time parked, and when parked, they were alone 99% of the time. When they were with the mother, infants spent 59% of the time in physical contact with her.

During the time spent with the mother, the infants spent 45% of time resting, 18% being groomed, 12% climbing, 7% jumping, 7% carried by the mother, 6% playing and 5% suckling. When parked the tarsier infants were mainly resting (93% of time), with jumping (5%), climbing (1%) and play (1%) occurring occasionally. The duration of behaviours varied: the mean duration for carries was 1.3 s (standard deviation (SD) = 1.7), 2.6 s (SD = 2.5) for climbing, 2.9 s (SD = 3.3) for play, 3.3 s (SD = 3.3) for grooming, 4.2 s (SD = 8.5) for jumping, 9.1 s (SD = 9.3) for suckling and 18.1 s (SD = 43) for resting.

3.2. Play behaviour

I observed locomotor, social and object play by both infants (I and II) during the study period. The infants started to perform play behaviour at the age of 2 weeks. Locomotor play involved climbing up and down on a branch and jumping for short distances back and forth, both when in proximity of the mother and whilst parked. Object play involved manipulating leaves or branches with hands or mouth. Social play could involve one-sided actions by the infant or in mutual play fighting. One-sided play occurred when the mother attempted to groom the infant, and it responded by touching and mouthing her head, back or tail. In addition, in situations when the mother was not engaged with the infant but was moving about in which case, the infant also touched and mouthed her tail and back whilst climbing on her. Mutual play fighting was observed later in the ontogeny. In infant I it was first observed at the age of 50 days. The mother was jumping around the canopy and apparently waited on landing at each new location for the infant, who was following her. Once the infant reached her, it grasped her back, and she responded leading to a mutual play fight that lasted for about 5 s. Mother turned back to the infant; both animals were grabbing each other by hands and mouthing/biting face to face. Then the mother turned away. Infant grabbed her back, but she did not respond anymore. Following termination of contact, the mother resumed her locomotion around the canopy. In infant II play fighting was first observed at the age of 34 days. Over a period of 6 min, I observed four events of play fight each several seconds long. As in the case with infant I, these play sessions occurred when the mother and infant were moving from the sleeping site to a new location and involved mutual grabbing and biting (Video 1).

http://dx.doi.org/10.5772/intechopen.70512

3.3. Maternal transportation of infants and the mother-offspring bond

Tarsier locomotion involves climbing and vertical jumping, and during the first days after birth, the mothers transported the infants. The mothers used their teeth to pick up the infants by the skin on their backs when transporting them. The mother usually held infant by the skin on the back but also on the side or even belly in one occasion (which was not comfortable so the mother then changed the position). The infant did not hold on to the mother baby when being carried, but on several occasions, an infant was observed to wrap its tail around the mother's upper back and neck, which may have provided some additional support (**Figure 3**). On other occasions the infant's tail passively hanged downwards. On completion of the transportation, the infant was left on a parking spot, and the mother went foraging. Parking lasted from as little as 4 to as much as 235 min (mean = 112 min, SD = 97.6). On her return, the mother would feed the infant. She would stay with the infant between 6 and 138 min (mean = 57.6 min, SD = 49.7), then transported and parked it again on a different location. On average infants were parked 162 cm (range: 30–400 cm), and during the study, tarsier infants were parked throughout the female's home range.

The infants started climbing branches at the age of one week and began to follow the mother over shorter distances between the third and fourth weeks (infant I at 18, infant II at 23 days). The mothers appeared to encourage the infants to follow by emitting chirping vocalization [22]. For longer distances (several metres) infants were still transported by the mother. By the age of 50 days, when observations ceased on infant I, it was no longer being carried by the mother. Observations on infant II ended when it was 35 days old, and at this age, it was still being carried.

By the time observations ceased on infants I and II, they were both still sleeping with their mothers. Infant (juvenile) IV was observed sleeping with his mother up to 10 months, infant V up to almost 9 months, infant VI at least up to 7 months and infant VII at least for 10 months. The inter-birth interval for the mother of infants IV–VIII was around 1 year.

3.4. Acoustic communication

Acoustic communication between mother and infant was very frequent and included four different signals out of eight signals described in Philippine tarsier (see Ref. [22] for detailed descriptions and spectrograms):

Loud call: a single note vocalization, capable of being carried relatively far, able to be reliably heard and identified from distance of at least 50 m. The loud call note consists of two distinct parts: the *whistle* part with a limited frequency range and the final *smack* part with a substantial frequency range. In adults the loud call was recorded mainly around sunset in proximity to the sleeping site but also during the night. In mother-infant communication, it was apparently used to gain others' attention during the night or when the mother was returning to the parked infant.

Chirp: a soft call that was heard mainly in proximity to the sleeping site in the evening or in the morning. Chirp consists of several short, rapid, broadband notes in a descending series. It was the most frequent call used by both mother and infant, with the mother using it enticing the infant to follow her. Chirps were also emitted by a mother over several nights after she had lost her month-old infant [19].

Cheep: is a single note call, usually emitted repeatedly in a sequence. It was recorded from both a mother and her offspring (infant III) whilst being rescued from poachers. It was also recorded in a free-ranging female (mother II). In this case the cheep was interspersed in a sequence of other call types.

Twitter was recorded during communication between a mother and her offspring (infant III) during being rescued from poachers. The signal consists of two parts, an initial long cheeping sound followed by several short, frequency-modulated twittering notes.

4. Discussion

Given the limited time available to collect data on the two main focal infants and the limitations imposed by attempting to make observations in difficult terrain and dense vegetation for all subjects studied, there are obvious deficiencies in the present study. Nonetheless, given the paucity in our knowledge of tarsier ecology and behaviour, I believe that the findings from the present study provide some important clues that show both similarities across species of the family and differences that may reflect variation in socioecology.

Births in the current study of Philippine tarsiers occurred between mid-April and mid-May, a pattern consistent with previous observations on this species [11] and also in spectral tarsiers [12], although, another study of spectral tarsiers reported two birthing seasons – November/December and May [2]. Records of births and the available literature on gestation length, 191 for wild spectral tarsier, 178 days for captive western tarsier and Philippine tarsier [17, 23, 24], indicate that the mating season is between October and January. Changes in testicle volume suggest a mating season between September and November in Philippine tarsier and from August to January in western tarsier [25]. A pair of captive Philippine tarsiers was observed mating at the end of October 2015 and beginning of November 2016 at the Tarsius Project Conservation Centre in Bilar, Bohol. The female then gave birth at the beginning of May after 188 days, 172 days, respectively (in the latter case, the former mating might be overseen) [pers. observ.].

Philippine tarsiers neither built nests or seek and use tree holes for giving birth. The observations in the present study showed that prior to giving birth, females either stayed high in the trees or remained close to the ground. This difference was most likely an artefact as the semi-captive female moved up the tree to avoid visitors who entered the enclosure every day between 9:00 and 17:00.

At birth, the infant Philippine tarsiers were fully furred, and their eyes were open. For the first month, or so, the infants were transported orally by the mother. The infants did not cling to her fur, but as apparent from **Figure 3**, they could help support themselves with their tail when being carried. Gradually over the first 3 weeks, the infants followed the mother for short distances whilst still being carried over longer distances. The mothers actively encouraged following by chirping. Of the two main forms of locomotion in tarsiers [9], infants began climbing up and down branches in the first week and jumping short distances by the second week.

Comparison of the limited data on Philippine tarsiers with the better studied spectral tarsier is instructive. The main difference between these two species is in their pattern of sociality and consequences this has on the care provided to infants. In spectral tarsier, which lives in extended family units, care of infants is not limited to the mother, but can also be provided by other members of the family unit [2, 3]. A major contribution to infant care is that members of the family unit, other than the mother, can regularly visit infants whilst they are parked. This babysitting or guarding afforded by other members of the family may enable mothers to engage in longer periods of unfettered foraging. In contrast, Philippine tarsier mothers are strictly solitary, and so all infant care is provided solely by them. Indeed, in only three occasions was an adult, likely males, encroach within 5 m of a mother-infant pair, and on two occasions the intruding individual was immediately chased away by the mother, perhaps to prevent infanticide. In one case, in spectral tarsier, a neighbouring male was observed to kill an infant [26]. Although, in another solitary species, western tarsier, males were often captured in close proximity to infants suggesting that fathers may provide some allocare. Even so, whilst male western tarsiers often visit the area around the sleeping site of young infants, they have never been observed to provide any form of parental care [27], and in captivity, males of this species have been observed to be aggressive towards other individuals including infants [17]. Therefore, it remains an unresolved issue as to whether solitary species like Philippine and western tarsiers have opportunities for fathers to provide allocare or whether all adult males are a potential risk for infants and so provide a constraint on how long mothers can leave their young parked and unattended. Infants do seem to have the opportunity to stay with their mothers for a protracted period. For example, our observations of mother-offspring pairs IV–VIII indicate that Philippine tarsier offspring can remain with their mothers until shortly before the offspring from the next year is born. Also consistent with such prolonged mother-infant association is a study that found that for three females, the young from the previous year remained even after the young from the next year were born [28].

In the more gregarious species, spectral tarsier, infants spent between 39% (in the first month) and 50% (in the third month) of night time alone; otherwise they are in close proximity (less than 5 m) of the mother or another group member (i.e., an adult male or subadults of both sexes) [3]. On the contrary, the mainly solitary Philippine tarsier's infants spent 66% of observed time parked alone. Also there are differences in the bout lengths of parking, with spectral tarsier the mean length of a bout of parking being 27 min (SD = 18.12, range 1–124 min), whereas my findings on Philippine tarsier indicate that bouts of parking are longer, with a mean of 112 min (SD = 97.6, range 4–235 min). If unrelated adult males pose a threat to infants, then the greater amount of time Philippine tarsier infants are left alone would place them at greater risk. Differences in habitat usage between social and solitary species may mitigate these risks, but such possible socioecological differences remain unknown (**Table 2**).

Consistent with risks to infants when parked, not only from potentially infanticidal males but also from predators, when parked the infants mainly remained at rest (93% of time), with little time spent moving about. In contrast, during the period spent with mother, the infants spent only 45% of the time resting. When parked alone, infants only spent 7% of their time jumping,

	Tarsius syrichta	*Tarsius spectrum*	*Tarsius bancanus*
Births	April–May	May (or November/December)	
Gestation length	172–188	191	178
Nest	No	Tree hole	No
Sociality	Solitary	Family groups	Solitary
Care of infants	Mother	Mother and other family members	Mother
Time spent alone during the first 2–3 months	66%	39–50% of time	
Mean length of a bout of parking	112 min	27 min	

Table 2. Comparison of data about reproductive behavior of three tarsier species.

climbing and playing, whereas when with the mother, 25% of their time was engaged these activities. Obviously, when with the mother, there was also the opportunity to engage in social behaviour (e.g., grooming, being carried and suckling), and these made up the remaining time spent being active.

Other than some preliminary reports on play of spectral and captive western tarsiers [2, 12, 17], there is little detail available on the play in tarsiers. The present study contributes to filling this void, by providing repeated observations of play in Philippine tarsiers. All three, commonly noted types of play—locomotor (solitary), object and social play [21]—were present in the Philippine tarsiers. The locomotor and the object play began earlier in ontogeny than social play. Locomotor play involved infants jumping and climbing around surrounding branches, and object play involved grasping, pulling and chewing leaves and small twigs. Social play, given that all other tarsiers were excluded from their vicinity, necessarily was restricted to infants engaging the mother, and this could take one of two forms. There was one-sided play, in which the infant grasped and mouthed the mother, who did not respond and remained passive or continued with ongoing behaviour, such as climbing. In play fighting the interactions were two sided, with both partners grasping, mouthing and grappling with one another. Similarly, in captive western tarsiers, these grappling bouts were characterized by the infant hopping at the mothers head and face with hands outstretched and mouth agape, followed by a short tussle in which mother and infant mutually mouthed and pawed at one another [17]. This pattern contrasts in many ways to the more gregarious spectral tarsier. In this species social play appears to be more frequent and easily observed [pers. observ.], and this may in part be because the infant has several other group members, not just the mother, as potential play partners. Indeed, mothers do not seem to play with infants, and although adult males do, it is less frequent compared to that involves the subadults, especially the females [12]. The differences in opportunity for social play may also account for the apparent species differences in the age of onset of social play: starting around the third week in spectral tarsiers and between 4 and 6 weeks in Philippine tarsiers. Similarly, social play reported in spectral tarsiers includes chasing and play fighting ([2], pers. observ.), whereas in the Philippine tarsiers and western tarsiers [17], it is mostly limited to short bouts of grappling. More detailed comparisons of the developmental onset, as well as the frequency and content of social play across a wider range of solitary and gregarious species of tarsiers, are needed to fully evaluate the causes of these species differences. Moreover, given that the social play in some primate species with more complex social systems can vary in marked ways with small differences in social systems (e.g., [29–32]), the possibility that even the small differences in degree of gregariousness can lead to significant differences in social play seen in tarsiers could provide novel insights into the evolution of such play [21].

The present study provides new information on the role of acoustic communication in mother-infant pairs. Mothers periodically transport infants to new locations, park them and the move around the neighbouring vicinity to forage. During these periods when the infants are parked alone, both the mother and infant emit vocalizations. The frequency of such calling appears to be much higher than in comparable situations when adults are foraging on their own with having an infant nearby. The most frequently emitted calls were chirps and less often loud calls. These two calls are also emitted outside the context of a mother-infant pair, suggesting that these signals may be used in a variety of contexts. Two signals, cheep and twitter, were only recorded during mother-infant pairs, suggesting that these have a unique role among such pairs. Cheep was recorded in rescued animals that were kept in a small cage before being released and may be a kind of distress call. However, it was also recorded in a free-ranging mother, who emitted this call in sequence along with other call types. The contextual relevance of all these calls needs further study to understand when they are emitted and why.

This present study also provides valuable information on the life history of Philippine tarsiers. Of the six infants observed, one was predated, representing a mortality rate of 17% during the first months of age. It is interesting that the predation occurred during the daytime when the infant was with its mother [19] rather than during night, when parked alone, when it would be expected that infants would be more vulnerable (see above). Although it may also be the case that during the daytime the tarsiers are sleeping and so likely less attentive to the presence of potential predators, unfortunately, data on mortality rates during day versus night are not available. A female Philippine tarsier can have (and raise) an offspring every year, as indicated by the five infants produced consecutively in one female during the present study. The inter-birth interval was around 1 year. The minimum inter-birth interval reported was 187 days in a captive female [24]; however, this infant did not survive, perhaps indicating that this may be shorter than the optimum.

In conclusion, the present study provides some preliminary, but important, data about Philippine tarsier life history and behaviour. These tarsiers can have one offspring per year, with an 83% chance of survival. Maternal care was documented to follow the cache and carry strategy with the infant spending around 66% parked alone whilst the mother was foraging. In this solitary species, infant care was provided solely by the mother. During the first days and weeks after birth, the mother transported their infants when moving to new locations. Whereas the infants were observed to start climbing on branches at the age of 1 week, they were transported orally until 3 weeks of age. Even though they started to follow the mother over shorter distances from three weeks, for longer distances, they were still transported by the mother until 50 days of age. Time budget data showed that when with the mother, infants spent 45% of their time resting, 18% grooming, 12% climbing, 7% jumping, 7% being carried by the mother, 6% playing and 5% of time suckling. Both infants engaged in locomotor, social and object play. Acoustic communication was very frequent between mothers and infants and included four different signals. Infants were observed sleeping next to mother up to 10 months, which is consistent with an inter-birth interval of 1 year that seems to be the case from study.

Acknowledgements

Special thanks are given to all co-workers helping with the Tarsius project especially Cristy Burlace and Václav Řehák for overall support; Lubomír Peške for help with radio-collaring; Filip Wojciechowski, Alex Wielbass, Sara Garau, Ema Knotková, Pavel Hrouzek and Phoenix Beamish for help with tarsier infant observation; and Carlito Pizzaras and Julius Baslot for help with locating the tarsier. Many thanks are given to Petr Slavík for providing the picture. I am thankful to the following supporting agencies Decin Zoo, Ústí Zoo, Primate Conservation, Inc, *Prosimian TAG of EAZA, WAZA* and *the Embassy of Czech Republic*, and the following local partners Wings of Serenity, The Philippine Tarsier Foundation, Inc. and Bohol Island State University. I thank the DENR Philippines for providing the permits necessary for the research. I am also thankful to the NIMBioS Working Group on Play, Evolution, and Sociality, at the National Institute for Mathematical and Biological Synthesis and the University of Tennessee, Knoxville, for encouraging me to publish data about play behaviour in tarsiers and especially to its members Sergio Pellis and Elisabetta Palagi for comments on early version of the manuscript.

Author details

Milada Řeháková

Address all correspondence to: tarsiusproject@gmail.com

Tarsius, z.s., Děčín, Czech Republic

References

[1] Perelman P, Johnson WE, Roos C, Seáanez HN, Horvath JE, Moreira MAM, Kessing B, Pontius J, Roelke M, Rumpler Y, Schneider MPC, Silva A, O'Brien SJ, Pecon-Slattery J. A molecular phylogeny of living primates. PLoS Genetics. 2011;**7**:e1001342

[2] MacKinnon JR, MacKinnon KS. The behavior of wild spectral tarsiers. International Journal of Primatology. 1980;**1**:361-379

[3] Gursky SL. Infant care in the spectral tarsier (*Tarsius spectrum*) Sulawesi, Indonesia. International Journal of Primatology. 1994;**15**:843-853

[4] Gursky S. Group size and composition in the spectral tarsier, *Tarsius spectrum*: Implication for social structure. Tropical Biodiversity. 1995;**3**:57-62

[5] Merker S. Habitat-specific ranging patterns of Dian's tarsiers (*Tarsius dianae*) as revealed by radiotracking. American Journal of Primatology. 2006;**68**:111-125

[6] Merker S, Groves CP. *Tarsius lariang*: A new primate species from western central Sulawesi. International Journal of Primatology. 2006;**27**:465-485

[7] Driller C, Perwitasari-Farajallah D, Zischler H, Merker S. The social system of Lariang tarsiers (*Tarsius lariang*) as revealed by genetic analyses. International Journal of Primatology. 2009;**30**:267-281

[8] Crompton RH, Andau PM. Ranging, activity rhythms and sociality in free ranging *Tarsius bancanus*: A preliminary report. International Journal of Primatology. 1987;**8**:43-71

[9] Dagosto M, Gebo DL. A preliminary study of the Philippine tarsier (*Tarsius syrichta*) in Leyte. American Journal of Physical Anthropology. 1997;**73**:1

[10] Dagosto M, Gebo DL, Dolino C. Positional Behavior and Social Organization of the Philippine Tarsier (*Tarsius syrichta*). Primates. 2001;**42**:233-243

[11] Neri-Arboleda I, Stott P, Arboleda NP. Home ranges, spatial movements and habitat associations of the Philippine tarsier (*Tarsius syrichta*) in Corella, Bohol. Journal of Zoology. 2002;**257**:387-402

[12] Gursky S. Allocare in a nocturnal primate: Data on the spectral tarsier, *Tarsius spectrum*. Folia Primatologica. 2000;**71**:39-54

[13] Gursky S. The behavioral ecology of the spectral tarsier, *Tarsius spectrum*. Evolutionary Anthropology. 2002;**11**:226-234

[14] Ross C. Park or ride? Evolution of infant carrying in primates. International Journal of Primatology. 2001;**22**:749-771

[15] Pellis SM, Pellis VC. The Playful Brain. Venturing to the Limits of Neuroscience. Oxford, UK: Oneworld Press; 2009

[16] Palagi E, Norscia I. The missing Lemur Link: An Ancestral Step in Human Evolution. Cambridge, UK: Cambridge University Press; 2016

[17] Roberts M. Growth, development, and parental care in the western tarsier (*Tarsius bancanus*) in captivity: Evidence for a "slow" life-history and nonmonogamous mating system. International Journal of Primatology. 1994;**15**:1-28

[18] Jachowski DS, Pizzaras C. Introducing an innovative semi-captive environment for the Philippine Tarsier (*Tarsius syrichta*). Zoo Biology. 2005;**24**:101-109

[19] Řeháková-Petrů M, Peške L, Daněk T. Predation on a wild Philippine tarsier (*Tarsius syrichta*). Acta Ethologica. 2012;**15**:217-220. DOI: 10.1007/s10211-011-0096-7

[20] Altmann J. Observational study of behavior: Sampling methods. Behavior. 1974;**49**: 227-267

[21] Burghardt GM. The Genesis of Animal Play: Testing the Limits. Cambridge, Massatchussets, London, England: MIT Press; 2005

[22] Řeháková-Petrů M, Policht R, Peške L. Acoustic repertoire of the Philippine tarsier (*Tarsius syrichta*) and individual variation of long distance calls. International Journal of Zoology. 2012;**2012**:1-10. DOI: 10.1155/2012/602401

[23] Izard M, Wright P, Simons E. Gestation length in *Tarsius bancanus*. American Journal of Physical Anthropology. 1985;**9**:327-331

[24] Haring D, Wright P. Hand-raising a Philippine tarsier, *Tarsius syrichta*. Zoo Biology. 1989;**8**:265-274

[25] Wright PC, Pochron ST, Haring DH, Simons EL. Can we predict seasonal bahavior and social organization from sexual dimorphism and testes measurements? In: Wright PC, Simons EL, Gursky S, editors. Tarsiers: Past, Present, and Future. New Brunswick. New Jersey, and London: Rutgers University Press; 2003. p. 260-273

[26] Gursky-Doyen SL. Infanticide by a male spectral tarsier (*Tarsius spectrum*). Primates. 2011;**52**:385-389

[27] Crompton, pers. comm. in 12

[28] Wharton CH. The tarsier in captivity. Journal of Mammalogy. 1950;**31**:260-268

[29] Palagi E. Social play in bonobos (*Pan paniscus*) and chimpanzees (*Pan troglodytes*): Implications for natural social systems and interindividual relationships. American Journal of Physical Anthropology. 2006;**129**:418-426

[30] Ciani F, Dall'Olio S, Stanyon R, Palagi E. Social tolerance and adult play in macaque societies: A comparison with different human cultures. Animal Behaviour. 2012;**84**:1313-1322

[31] Palagi E, Cordoni G. The right time to happen: Play developmental divergence in the two Pan species. PloS One. 2012;**7**:e52767

[32] Reinhart CJ, Pellis VC, Thierry B, Gauthier CA, VanderLaan DP, Vasey PL, Pellis SM. Targets and tactics of play fighting: competitive versus cooperative styles of play in Japanese and Tonkean macaques. International Journal of Comparative Psychology. 2010;**23**:166-200

Chapter 4

Approaching Human Dimensions in Lemur Conservation at Lake Alaotra, Madagascar

Lena M. Reibelt and Patrick O. Waeber

Additional information is available at the end of the chapter

http://dx.doi.org/10.5772/intechopen.73129

Abstract

'Human dimensions of wildlife management' is a concept that emerged some 50 years ago and has gained global application. A majority of cases report on human-wildlife conflicts (HWCs), where wildlife is causing problems to an expanding human population or vice versa. In Madagascar, lemurs represent a flagship for conservation. Many lemur taxa are threatened, and conservation is facing increasing challenges due to habitat loss and degradation. The Alaotran gentle lemur (*Hapalemur alaotrensis*) is the only marshland living lemur. Its conservation is particularly challenging due to various conflicting interests of different stakeholder groups. The Alaotra region is the bread basket of Madagascar, producing a majority of inland fish and rice. Here we present a new venue taken by conservation, which is based on a transdisciplinary research approach, participatory modeling, and gaming through role-playing games (RPGs). This holds promise to engage stakeholders from the onset of conservation planning and management, and it is hoped that increased participation will spur ownership and thus reduce conflicts among stakeholders to increase conservation effectiveness to safe *Hapalemur alaotrensis* from extinction.

Keywords: *Hapalemur alaotrensis*, human dimensions of wildlife management, human-wildlife conflict, transdisciplinary research, participatory modeling, role-playing games, stakeholders

1. Introduction

Human dimensions of wildlife conservation evolved in the 1960s and gained increasing attention in the past 30 years, both in research and among practitioners. While biological and ecological paradigms dominated natural resource and wildlife management for long, increasing human-wildlife conflicts spurred the awareness that people aspects need to be included in management decisions as they are critical to conservation success [1, 2]. 'Human dimensions' is a broad field today, which concerns the question how to best manage wildlife, that is, how to ensure species' survival without compromising people's needs. It occurs in as diverse settings as agriculture, hunting, tourism, and leisure realms and entails efforts to understand and affect human behavior by incorporating insights about people's attitudes, perceptions, and norms into policy and management programs [2, 3]. Increasing overlap and interference in landuse caused increasing incidents of human-wildlife conflicts in the past few centuries. Such conflicts may occur when elephants raid farmers' fields, when wolves predate on domestic sheep, or when humans shoot a lion that is thought to have killed cattle. Newer definitions also incorporate human-human conflicts evolving from human-wildlife conflict, for example, individuals being negatively affected by wildlife versus conservation organizations or state authorities [4, 5].

Per definition, human-wildlife conflict (HWC) occurs when "the needs and behavior of wildlife impact negatively on the goals of humans or when the goals of humans negatively impact the needs of wildlife" (World Park Congress Recommendation as cited in [4]). Human population growth coming along with land reclamation and cultivation in formerly uninhabited areas is one of the main reasons for increasing human-wildlife conflicts [6]. HWC occurs globally and concerns a variety of species and sociocultural and socioeconomic contexts, including mammals, fish, insects, and reptiles globally. The range of human-wildlife conflict includes lions, monkeys, and elephants in Africa, leopards and tigers in India, or wolves in Canada, USA, and Europe, to name but the most prominent ones [7].

Regardless of the HWC context, some main characteristics do apply. For example, communities are not homogeneous entities, but incorporate different stakeholder groups with different needs and value systems [8]. Incorporating these different views and finding acceptable 'solutions' for all parties involved and affected is a complex and complicated task for conservation management and planning. With increasing recognition about the importance of the human dimension, stakeholder participation became more important; moreover, research collected evidence that conservation projects are more likely to be successful if locals are involved in management decisions and conservation planning [9]. Decker and Chase [3] identified five main approaches how wildlife managers can seek public participation. They differ in the degree of influence of wildlife managers and stakeholders on policy and management decisions, beginning with highest influence of the managers and lowest of stakeholders with the (top-down) authoritative approach. With decreasing own influence, wildlife managers can increasingly let stakeholders contribute to decision-making with the passive-receptive, inquisitive, transactional, and co-managerial or delegation approach (cf. Figure 1 in [3]). Current literature suggests that management decisions and rules tend to be better accepted when stakeholders were involved in the decision-making process. One reason for this is that attitudes, aspirations, and norms are better understood and can be incorporated in decision-making. In order to resolve or alleviate human-wildlife conflict, human dimensions thus encompass people's beliefs, values, attitudes, behaviors, and socioeconomic and demographic characteristics of individual stakeholders or stakeholder groups; it deals with the proximate level of interaction among and between management decisions, processes, and staff (cf. [10]).

http://dx.doi.org/10.5772/intechopen.73129

Primates represent a particular case in the HWC realms. While they are similar to humans and venerated in some settings, people perceive them as pests in other instances, while the contexts are ranging from agricultural fields to reserves and tourist camps to towns [11]. Major threats to primate populations are conversions of natural habitat into areas of human use such as forestry, plantations, and agricultural fields; trapping for biomedical trade, bushmeat trade, and transmission of diseases represent further threats [11, 12]. While HWC concerning primates such as baboons, vervets, and macaques is well covered in the scientific literature (e.g., [11, 13–16]), Madagascar's lemurs are hardly considered even if the majority are endangered and efficient management measures are needed to halt further population declines.

1.1. Lemur conservation in Madagascar

Madagascar hosts a unique assembly of endemic primates. The lemurs are a monophyletic group of strepsirhine primates occurring only on Madagascar [17] consisting of five families: Daubentoniidae (1 species), Indriidae (19 species), Lemuridae (21), Lepilemuridae (26), and Cheirogalidae (36) [18]. In the last 10–15 years, advances in molecular biology have resulted in an increase from some 50 to 107 known lemur taxa [19, 20]. Lemurs exist in nearly all of Madagascar's forest ecosystems, from the very dry spiny forests of Madagascar's southwest, along the dry forests of the west, and along the entire east coast in the subhumid and humid forests [20, 21]. All lemurs are nationally and internationally protected species. Ordinance No. 60-126 of October 3, 1960, represents the first official national text on the protection and hunting regulations of wild species, including lemurs (cf. [22]). Madagascar signed the CITES (Convention on International Trade in Endangered Species of Wild Fauna and Flora, also known as the Washington Convention) in 1975 and added all lemurs shortly thereafter to CITES Appendix I (Decree 77–276 of August 26, 1980). To protect its unique wildlife, Madagascar was among the first countries to establish a protected area network of National Parks with the first created in 1927. In the Durban Vision proclaimed in 2003 during the Fifth World Parks Congress in South Africa, then President Marc Ravalomanana declared to triple the terrestrial surface in Madagascar up to some 10% of Madagascar's land under some sort of formal protection [23, 24]. Despite all these formal agreements and laws, many lemur species are threatened by habitat loss and hunting (for both bushmeat and pet trade; [25–28]). The 2012 IUCN Red List evaluation of the threat status of Madagascar's lemurs shows that 94% of species are either classified as Vulnerable, Endangered, or Critically Endangered [29]. The biggest challenge to lemur conservation is the fast pace of deforestation with agricultural production and infrastructure being two of the main proximate drivers [30], thus leaving many lemurs in isolated forest fragments [31]. Forest fragments are highly susceptible to anthropogenic change and thus some lemur populations and species, even some of the newly discovered species risk disappearing due to their lowered resilience in fragmented or degraded habitats (e.g., decreasing numbers, loss of genetic diversity, increased disturbance [32]). Degraded forests are furthering the exposure of primates to humans [33], and their close phylogenetic relationship puts them more at risk of disease transmissions from humans to primates [34]. In Madagascar, it has been shown that *Avahi laniger*, *Eulemur rubriventer*,

Hapalemur aureus, Microcebus rufus, Propithecus edwardsi, and *Prolemur simus* have increased diarrhea cases due to exposure with human enterobacterium [35]. While bushmeat represents a major threat for Madagascar's lemurs, a traditional form of taboos, called *fady,* protects some lemur species from hunting and consumption. The *fady* largely are ancestral rules which are still respected by a majority of the Malagasy people; however, these taboos differ from region to region, and with increasing mobility, an increasing number of taboos are weakened by immigrants from other tribes and regions (for more details, see [36–38] and references therein). In this context, the abandonment of an old attitude or tradition can cause significant conservation issues when exploitation suddenly is no longer seen as socially inacceptable. For example, the *fady* formerly protecting *Indri indri* and *Propithecus verreauxi* from consumption are less respected today [36]. However, *fady* can also cause increased hunting pressure. The aye-aye (*Daubentonia madagascariensis*), for example, is oftentimes killed when encountered because the nocturnal lemur is believed being an evil omen bringing disease or death to family members or even whole villages if encountered and not killed [39].

Opposed to the situation in many other contexts, crop-raiding evidence in Madagascar is scarce. Still, some species have been reported to raid, for example, *Propithecus verreauxi coquereli, Lemur fulvus, Avahi laniger occidentalis,* and *Leplilemur mustelinus edwardsi* targeting cashew fruit, mango, and tamarinds [11, 40], but also *Daubentonia madagascariensis* [41]. In general, HWC in Madagascar is mainly represented by habitat loss of lemurs, while lemurs themselves do not directly affect farmers' subsistence. A key issue to conservation in Madagascar is the increasing competition between humans and wildlife. Human population growth results in spreading of human activity such as agriculture, precious wood, and stones into areas which were wildlife habitats before. With the protection of wildlife habitat (e.g., the establishment of protected areas), local people may be restricted in extending their fields or activities, or are even forced to relocate, thus encountering opportunity costs due to land-use restrictions or hunting bans [42–44]. A widely used approach to engage conservation with local resource users is community-based conservation (CBC). It has been designated to be the most practical approach to fight biodiversity loss in developing countries [45]. However, it has also been considered as time-consuming and complicated, and criticized that it does not necessarily provide win-win situations, but that losers may be generated through the transfer of rights, power, and resources as well [46–47]. The approach is oftentimes implemented for the promotion of development or livelihood security while reaching conservation goals as well [48]. The biggest challenge lies in the intrinsic complexity of the conservation and development issue(s), combined with the multitude of different contexts that makes simple upscaling or transfer to other sites literally impossible. Each case involves a multitude of different stakeholders and resources, different power relationships, and management priorities. These complex socioecological issues require the consideration of multiple perspectives, worldviews, and priorities.

A community-based conservation approach is also being implemented in the Lake Alaotra region (**Figure 1**) to preserve the Alaotran gentle lemur (*Hapalemur alaotrensis;* **Figure 2**) [49]. In the remaining of this chapter, we will present the case study of the Alaotra, Madagascar's rice granary. We will describe the Alaotran gentle lemur, its conservation challenges and analyze

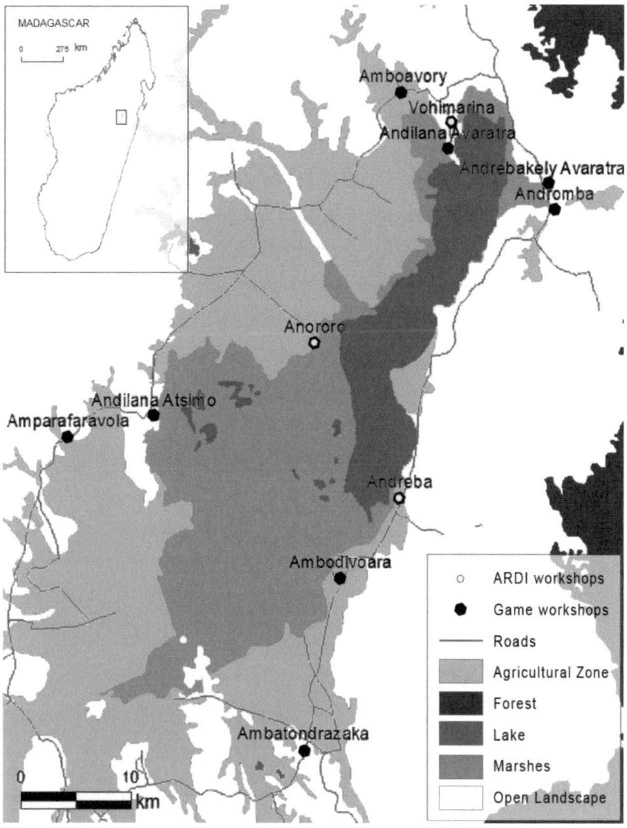

Figure 1. Lake Alaotra region. The map shows the lake, with surrounding marshes, rice fields, open landscapes (dominated by grasslands), and forests. Intervention villages are situated around the Alaotra wetland. This map has been modified from Reibelt et al.'s Figure 1 published in the Journal Madagascar Conservation & Development under a Creative Commons Attribution 3.0 Unported License.

the human dimensions of the conservation endeavors in the marshes and communities around Lake Alaotra.

1.2. The Alaotran gentle lemur and its conservation

The Alaotran gentle lemur (*Hapalemur alaotrensis*) is globally unique, living in and restricted to the marshlands of Lake Alaotra. It represents one of the five extant species in the genus *Hapalemur*. The other four species, namely the southern bamboo lemur (*H. meridionalis*), northern bamboo lemur (*H. occidentalis*), lesser bamboo lemur (*H. griseus*), and golden bamboo lemur (*H. aureus*), all are forest dwellers, occupying a variety of forest types across Madagascar. Genetically and phenotypically, *H. alaotrensis* is closely related to *H. griseus*, and it is hypothesized that the marshland living one must have originated from the forests before humans settled in the Alaotra some few hundred years ago [50–52]. *H. alaotrensis* is a crepuscular primate and performs cathemeral activity behavior [53], shows female dominance, a

Figure 2. A juvenile *Hapalemur alaotrensis,* with permission from photographer Arnaud De Grave, Le Pictorium Agency.

social behavioral trait common in many lemurs [54], and has a specialized diet based on marshland vegetation only [53] and it is well adapted to wetland conditions [51]. *Hapalemur alaotrensis* is the only primate species in the world that lives exclusively in a wetland habitat. The species is classified as Critically Endangered [55] due to its very restricted geographic range. *Hapalemur alaotrensis* is at high risk of extinction due to rapid and ongoing habitat destruction for conversion of the marsh to rice fields [56]. The marshland coverage was around 19,000 hectares in the early 2000s and it has decreased to below 14,000 in the mid-2000s; in 2012, there was an extreme fire year affecting more than 50% of the remaining marshes [57]. There are two factors leading to increased marshland burning: lack of law enforcement and prolonged drought seasons. For example, a single rice field was found within the Park Bandro at Andreba in 2013, but this increased to five rice fields in 2014. People in Andreba stated that they will transform the marsh into rice fields if the current delinquents are not punished. A census of Madagascar Wildlife Conservation (MWC), a Malagasy NGO, revealed that in 2016, a fourth of the park was covered with illegal rice plantations [58].

Lemur conservation in the Alaotra is intricately complicated and complex constituting a typical wicked problem[1] (sensu [59]) as many conservation problems (see also [60]). As is typical for wicked problems, there are a multitude of stakeholders involved in the Alaotra region who are directly or indirectly linked with the wetlands (cf. [61]), each with their own worldviews, values and knowledge systems, ending up having divergent and sometimes opposing or even conflicting interests or agendas. There are several different governing institutions that sometimes share overlapping responsibilities and tasks; there are, for example, the Ministry of Environment that is responsible for the wetlands and forests and the Madagascar National Parks that are responsible for protected areas which sometimes fall on forests or as here on wetlands; there is the Ministry of Fisheries responsible for the overseeing of lake-wide activities or the Ministry of Agriculture governing all land-based activities that fall within the agricultural domain and the open landscapes and wetlands. There is a strong position for lemur and biodiversity conservation in general, since there are endemic species found in the Alaotra (e.g., *Hapalemur alaotrensis* and *Salanoia durrelli*). In addition, intact marshes have an important role for functioning ecosystem services such as water retention, filtering, and water quality [62]. On the opposing side, there is a strong lobby promoting the conversion of marshlands for rice production, since rice is a quality of life [63] and an important staple food in Madagascar, especially in the Alaotra [64]. Rich people from outside the Alaotra are interested in buying land for turning it into rice fields [65].

1.3. The human dimension in the Alaotra region

A startling issue among conservation biologists is the fact that conflict management oftentimes is tackled by making assumptions about human attitudes and behaviors which are seldom congruent with reality [5]. Research has shown, however, that conservation projects benefit by taking into account the needs, attitudes, and aspirations of locals in order to increase the efficacy of conservation efforts [5]. In the Alaotra, there is an immense anthropogenic pressure on biodiversity and the natural ecosystems. In order to strike a possible balance of biodiversity values with the growing need for agricultural products and other ecosystem services, the understanding of livelihood needs, the main resource users' attitudes toward and perception of life, livelihood, and lemurs become essential to inform conservation planning. Resource users who work in the marshes would prefer land sparing to land sharing, that is, having clear demarcation zones for work (e.g., fishing and farming) and zones for biodiversity conservation such as is the case for the special conservation zone Park Bandro [66]. People also seem to have a neutral or even positive attitude toward the lemurs and their conservation as long as they can pursue their livelihood activities [56]. Community members in general view the environment as a social construct or related to

[1]Wicked problems are characterized by the following attributes: (1) formalizing the problem is not possible, that is, the construction of a solution space is the core challenge since every problem is a symptom of yet another problem; (2) no evident stopping rule exists (when is it solved?); (3) there is no true or false solution to it, and rather, there is only a better or worse one; (4) every decision "will generate waves of consequences over an extended (…) period of time" ([59]: 163)—tracing all the consequences is impossible, especially when the half-lives of such are long; (5) every solution constitutes a "one-shot" operation where every implementation is consequential [59].

human benefits; for example, the introduction of the invasive snakehead fish (*Channa maculate)* is seen as positive by local teachers since it delivers additional proteins to people [67]. Stoudmann and colleagues [63] identified five main livelihood attitudes in the Alaotra region: (1) 'Responsibility makes a man' refers to people who take their situation into their own hand, and which are involved in improving the state of things. (2) 'Let us be realistic' is a rather fatalistic view of things by people who think that one cannot do much about life; (3) 'Children are the future,' an attitude shared by mainly women, who are concerned about changes affecting the next generation, who acknowledge the importance of education, and who are worried about teenage pregnancies. (4) 'Good things come to those who work hard' is the attitude shared by people who pursue various agricultural strategies (e.g., fertilizer, crop diversification), and who believe that working hard will improve their standard of living. (5) 'Be prepared for the unexpected' is a group of resource users who, similar to the previous group, invest heavily into diversification and who have an entrepreneurial spirit [63].

A majority of Alaotra's rural population are engaged in some sort of agricultural or fishing activities [68]. Livelihood conditions are becoming harder. For example, annual fish catches amounted to 4000 t in the 1960s, making the Alaotra the most important inland fishery region of the country. Recent numbers have been dropping to below 800 t per annum [69]. The steady decline of the fish stock is most likely a result of overfishing, acidification of the lake, introduced fish species, and siltation [70–73]. A majority of the marshlands fringing the lake have already been converted for rice production, with some 100,000 ha outputting ca. 300,000 t per year [57, 74]. However, changing environmental conditions (e.g., deforestation and clearing of surrounding hill slopes through slash-and-burn agriculture known as *tavy,* leading to increased siltation of affluent rivers) have diminished the lake size to 20% of its former size in 2000. Continued dry spells could soon let wither Lake Alaotra and make it another 'case of Lake Baikal'; according to Bakoariniaina et al. [75], some 5 km^2 of lake surface have disappeared within a period of 30 years. Consequently, rice crop productivity in the Alaotra basin has dropped to about 40% of its former level [75, 76]. The stakeholders' livelihood strategies for meeting their needs and to cope with changing socioeconomic and environmental conditions are currently resulting in a lose-lose scenario (sensu [77]) in the Alaotra socio-ecological system. The area is home to almost one million people, thus being the highest population density in a wetland area in Madagascar [78]. Arable land is becoming increasingly scarce, forcing many people into the marshes to establish rice fields [57, 66]. Extensification is still widely common compared to intensification [79]. Recent years experienced extended drought periods in the region, with some years receiving almost no rain at all. This showed two main consequences in the socioecological system of the Alaotra. (1) People have been pushing into the marshes and lake for establishing rice fields or claiming the territory for future rice production by staking and building so-called *hamatra* or reed fences. These have started to even crisscross the entire lake from east to west, thus casting a grim picture of the future Alaotra as being one big rice field. (2) The reduced water availability in marshes and lake has negatively affected the lake-wide fish stock. There is less and smaller fish available [73, 80, 81]. More people than ever are pushing into the marshes and lake to fall back on fishing as a last resort, sometimes even using mosquito nets in despair.

2. A transdisciplinary approach for lemur conservation

Current environmental destruction trends do not cast an optimistic future for the survival of *Hapalemur alaotrensis*. The unique wetland primate may disappear from the Alaotra and Madagascar within the coming years if habitat destruction cannot be halted. Concerted conservation efforts are underway engaging various NGOs working together with the riverine communities and the authorities. It is hoped that increased collaboration and the adoption of a transdisciplinary approach will allow pushing back the threats that *Hapalemur alaotrensis* and its habitat encounter. Activities and programs include habitat restoration, marsh patrols, various research projects [49], environmental education [67, 82], and awareness-raising campaigns including the World Lemur Festival (called Bandro Festival; based on the vernacular name of *Hapalemur alaotrensis*). However, these are not enough. Thus, a new conservation approach for the Alaotra is been unfold in order to slow down the main threat to the existence of the unique marsh lemur, the habitat destruction for agricultural production. Habitat restoration is a classic conservation activity and urgently needed in the Alaotra to link isolated subpopulations [49, 56]. Complex conservation problems require creative approaches [60]. What is creative about habitat restoration? MWC accompanies all its conservation efforts with games. The serious gaming approach, which requires intense exchange and communication with stakeholders, is based on Companion Modeling (ComMod, cf. [83]). It builds on an inductive process of creating conceptual models from field evidence and judgments with restitution to knowledge providers in the form of interactive games. The central tool is a model or game which can, depending on the conservation issue and needs at stake, be a research means to elicit different potential future scenarios and stakeholder responses to this, or a communication tool to bring different stakeholder groups together to exchange and discuss on possible management options. In such game settings, the resulting outcomes are shaped by the cumulative and sometimes interacting decisions made by individual players, coupled with all the interacting decisions by the other players, as well as the rules of the game. The games are strategic situations [84], thus representing effective tools for exchange and solution seeking in decision-making and scenario planning contexts (cf. [85]). There is a strong relationship between game behavior and players' real life (e.g., [86]) and this tool thus represents a valid alternative to more classic social science approaches.

This inter- and transdisciplinary research (sensu [87]) aims at understanding farmers' perceptions and attitudes and allows farmers to explore the ecological, economic, and sometimes social outcomes of their individual and cumulative decisions. At the onset of the ComMod approach are participatory workshops (field work), where stakeholders share their mental models or mind maps with each other and the researchers, that is, where group discussions describe and develop a common representation of the socioecological system which all involved parties can agree upon. The methodology is based on ARDI [88] and represents a dedicated participatory modeling method. Researchers and stakeholders identify the main Actors, Resources, Dynamics, and Interaction (ARDI) being relevant to the socio-ecological system at stake and the agreed-upon issue(s). Then, the identified components are translated into players, game components, and rules for a role-playing game.

2.1. The games

2.1.1. *A wetland game for research*

The wetland game [81] has been developed for research purposes to understand land-cover-type changes and livelihood strategies of the main resource users living around Lake Alaotra. The research processes and results are aimed to include the human dimensions into conservation planning. The developed wetland game is a role-playing game (RPG) consisting of a market, a landscape, and a bank. Players are farmers and fishers as in real life and can do fishing and different farming activities (rice, onion, and vegetables) and invest into technology (compost). They further have the choice to do opportunity activities (logging, mining, and hunting), invest into housing (three quality levels) or different quality of life parameters, namely protein, electricity, health, and education (**Figure 3**). Players track individual decisions on their personal player sheet and subsequently place their activity tokens on the game board, the common landscape (**Figure 4**), which represents the different land types in the Alaotra region (lake, agricultural zone, hilly grasslands, and forest (see [81] for details on the game development process)). Consequently, the common landscape shows the cumulative decisions and impacts of all players. These changes are mostly represented by changes in the original land cover type, which are induced by land-based activities (e.g., farming in the marshes transforms them into agricultural zone, which is indicated via color change of the respective cell(s) [81]. The bank (represented by a ComMod team member with a computer) tracks all the players' decisions and calculates and pays the cash output. The agricultural or fishery production depends on factors such as how many other resource users are sharing the same space, how much of the original land type is still intact, or how is the weather (i.e., is there a climatic event such as drought or cyclone). The researchers are accompanying the gaming phase by quantifying the activities, thus decisions taken by the players and the impacts on the common landscape. Moreover, a qualitative phase follows the gaming where experiences during the game are shared, discussed, and explored. It is during the debriefing [89] where there is room

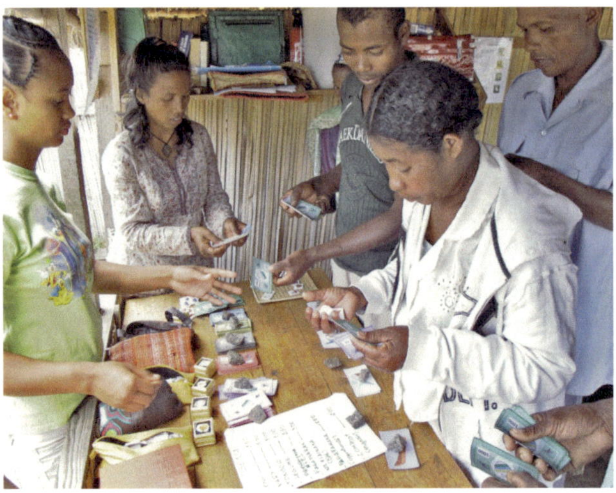

Figure 3. Wetland research game: players buy activities at the market.

Figure 4. Wetland research game: players track how their individual decisions accumulate on the common landscape.

for freeing emotions and understanding what happened during the game, in order to then bridge the virtual game reality with the real world. Important to note here is that game behavior does not necessarily reflect reality; the game behavior serves as entry point to compare game activities and real life. It also frees people of the social constraints that are often accompanying people during an interview or open discussions on topics of potential conflict. A follow-up monitoring 1 week after the workshop allows researchers and players to exchange one by one on further details and thoughts concerning the issue at stake. Participants acknowledged the opportunity to openly discuss land-use strategies and decisions and appreciated the fact that they could also exchange controversial ideas in the workshop setting without entering in disputes. Several gamers perceived that the game offered them a new, broader perspective on their surroundings and the ongoing processes in reality. They further described in the debriefing sessions that game behavior matched real-life behavior from about 50 to 100% and thought that the gaming experience would help them to make better decisions in the future.

2.1.2. A wetland game for discussion and outreach

"There is no right way to do conservation. There are only choices" [90]. Scales [91] adds that "To help make these choices, research and policy in Madagascar desperately need more conversations-between biologists, anthropologists, archeologists, economists, environmental historians, and geographers; between researchers and practitioners; and between 'experts' and the individuals, households and communities directly dependent on the island's natural resources for their livelihoods." Exchange is crucial for effective learning and to avoid repeating the same mistakes over and over again [48]. In order to have a game which can be used for exchange, negotiation, and outreach, the research game was simplified accordingly.

The discussion game is a follow-up of the wetland research game and was designed to develop, discuss and explore rules and regulations in the context of marshland conservation and management. The discussion and communication tool is a simple representation of the Alaotra socioecological system, representing the lake with fish, the marshes with biodiversity and the agricultural zone with rice fields (**Figure 5**). The basic rule is that all eight players need one fish token and two rice tokens at the end of each round to feed their family. Players are free to plant rice, use compost, and go fishing as they wish. As soon as a resource is depleted, or players do not have enough production to survive, the game is interrupted to discuss game behavior and consequences and, foremost, to elaborate on how the situation could be improved when replaying the game. Players are thus invited to establish game rules (e.g., restrictions on fish catch or prohibition to transform marsh) and to test the success of their established rules and whether these are suitable to reaching a sustainable system where everybody can survive (**Figure 6**).

The game calibration shows the most crucial linkages and interdependences in the Alaotra system. The marshes are breeding ground for some fish species; thus, reproduction reduces with shrinking marsh area (i.e., when players burn the marshes to establish rice fields). Moreover, the marshes play a role in water availability in the system. With each transformed marsh patch, there is less water available, which has a negative impact on rice output. Finally, there are less marsh patches than players available; this is a proxy for increasing population and the fact that there is not enough (marsh-) land available for everyone. The game thus addresses a common pool resource situation with the fish (and the marshes) and helps exploring the question of what future management scenarios are possible and which could be embraced by the communities? The game serves as a simple window to the future, helping the local stakeholders to become aware and understand current trends in the system and potential consequences of their decisions.

First results suggest that people tend to intensively (over)exploit the system if they have the opportunity to do so. The players quickly establish new effective game rules, which show high similarity to already existing conservation rules. The strength of the game is that the participants can discuss prerequisites, advantages, and disadvantages of different potential rules and then decide themselves which one to try out. During the testing workshop debriefings, participants emphasized the interdisciplinary nature of the game, its suitability for rural resource

Figure 5. Wetland discussion game: prototype representing lake, marsh, and agricultural zone.

Figure 6. Wetland discussion game: regional authorities implementing a self-developed rule during testing phase.

users but also school children, and judged the game to be realistic, instructional, enjoyable, and suitable to enter into fruitful discussions. It still remains to be tested whether this game approach can increase the acceptance of already existing conservation rules in the real Alaotra socioecological system.

2.2. Concluding remarks and outlook

Over the course of the past 5 years, the conservation community was able to substantially enhance its understanding of the human dimensions of *Hapalemur alaotrensis* conservation around Lake Alaotra. What are conservation management preferences for the subsistence farmers and fishers? What are their attitudes toward the endemic lemur or the core conservation zone Park Bandro at Andreba [56, 66]? What are the rural stakeholders' strategies to cope with change [63], and how do they take decisions in the agricultural domain [64, 81]? The next step is now to implement all the gained knowledge and understanding to enhance conservation actions and continue the dialog of trust with the different stakeholders. It is assumed that the intense exchanges and workshops enhanced understanding and respect on both sides, and this will be fundamental in the implementation phase. The deployment of role-playing games helps conservationists to engage with various stakeholder groups to spur discussions to increase knowledge and understanding of problems at hand. It helps the stakeholders to elicit their mental models and to strengthen their adaptive capacity and critical thinking, and foremost, it holds promise to strengthen their ownership in resource management and planning.

The recent research efforts highlight that local resource users are not basically against conservation of the marshes and its biodiversity; nevertheless, protected lemur habitat is shrinking constantly and at faster pace in recent times. Lemur habitat restoration measures are urgently needed, but these will likely interfere with peoples' newly established rice fields. Even if the fields are formally illegal since placed within the New Protected Area, law enforcement is weak, and people depend on their rice harvest. Restoration actions thus bear a high risk to increasing HWC in the sense of human-human conflict, that is, farmers versus conservationists. Here, it is critical to intensely exchange and communicate with the communities involved and to further integrate the various stakeholder groups in the decision-making process. Current plans by Durrell Wildlife Conservation Trust and Madagascar Wildlife Conservation (both have been active in the region for over 20 and 14 years, respectively) and other collaborating NGOs are to reconnect the isolated Park Bandro with marsh habitat and further lemur subpopulations in the south. It is critical where to reforest marshes in order to increase chances that the newly planted cyperus shoots will not be destroyed immediately by fishers or farmers who were not part of the decision process. Conservation bodies and researchers will thus organize planning workshops with the adjacent communities and involve all crucial stakeholders such as official and traditional village leaders, VOI (responsible entity for natural resource management), affected fishers and farmers. Based on ecological data, different scenarios will be developed and then discussed and assessed with the stakeholders to include the human dimensions, that is, their attitudes and preferences. The aim is to reconcile both human needs and biodiversity values.

The understanding obtained in the various meetings, workshops, and gaming sessions will help with this difficult task. There is seldom one 'solution' or 'answer' to conservation issues or human-wildlife conflict, but different choices, which are more or less acceptable to different stakeholders or interest groups [3, 8]. Stakeholder involvement and negotiation processes are crucial to determine acceptance of proposed management in advance [92]. In Madagascar, a disconnection of policy decisions and community needs has reduced the effectiveness in the conservation and development sector in the past 30 years [79]. Especially in poor countries, people sometimes feel as victims of top-down decisions in conservation, which impact their lives and livelihoods without giving them the opportunity to take part in the decision-making process. This can create resistance or opposition toward conservation projects and conservation organizations. Including local resource users in the conservation planning process creates feelings of ownership and increases chances of long-term success of conservation projects. This link may explain why the special conservation zone Park Bandro is still existent and well-respected by the majority of the adjacent community of Andreba because it was created together with the community (but see [49] for details).

As in the global conservation movement, initiatives for the protection of *Hapalemur alaotrensis* initially focused on habitat protection and ecological insights. However, with growing human pressures, the human dimension increasingly became more prominent in management decisions and conservation strategies. In the past century, conservation advocates realized a broad range of conservation and development projects, reaching from basic reforestation and exploitation regulations, over education and outreach initiatives, to agriculture support and improvement. However, with ever-increasing human population numbers (both local increase and immigration) and changing climatic conditions, it is becoming increasingly challenging to convince people of conservation importance. Weak law enforcement corrodes conservation success in many developing countries, especially when rural people can increase their little income through illegal activities [93]. Law enforcement is thus critical in protected area management to ensure long-term conservation success [94–96]. Though the integration of mutual benefits for human wellbeing and biodiversity has gained increasingly attention in Madagascar following global trends, the challenges of realizing this by community involvement and co-governance in Madagascar remain the same: the management and monitoring of these areas is proving to be difficult due to a combination of a lack of financial and human resources, as well as weak technical capacity [97]. This makes the human dimensions even more important; considering local resource users' needs and aspirations and including them in the decision-making process has been proven in many other contexts to increase ownership, support, and long-term conservation success. The fact that people in the Alaotra region are willing to negotiate conservation zones in the marshes raises hope that *Hapalemur alaotrensis*, currently being listed as 1 of the 25 most endangered primates in the world [98], still has chances of survival.

Acknowledgements

This chapter is based on work supported by the Margot Marsh Biodiversity Foundation under research grant PR15: 021, and the Swiss programme for research on global issues for development under research grant IZ01Z0_146852 as part of the AlaReLa Alaotra resilience landscape project.

Author details

Lena M. Reibelt[1*] and Patrick O. Waeber[1,2*]

*Address all correspondence to: reibelt.lena@gmail.com and patrick.waeber@usys.ethz.ch

1 Madagascar Wildlife Conservation, Ambatondrazaka, Madagascar

2 Ecosystems Management, Forest Management and Development Group, ETH Zurich, Zurich, Switzerland

References

[1] Decker DJ, Richmond ME. Managing people in an urban deer environment: The human dimensions challenges for managers. Urban Deer: A Manageable Resource. 1995:3-10

[2] Manfredo MJ, Dayer AA. Concepts for exploring the social aspects of human–wildlife conflict in a global context. Human Dimensions of Wildlife. 2004;**9**:1-20

[3] Decker DJ, Chase LC. Human dimensions of living with wildlife: A management challenge for the 21st century. Wildlife Society Bulletin (1973–2006). 1997;**25**:788-795

[4] Madden F. Creating coexistence between humans and wildlife: Global perspectives on local efforts to address human–wildlife conflict. Human Dimensions of Wildlife. 2004;**9**: 247-257

[5] Dickman AJ. Complexities of conflict: The importance of considering social factors for effectively resolving human–wildlife conflict. Animal Conservation. 2010;**13**:458-466

[6] Woodroffe R, Thirgood S, Rabinowitz A, editors. People and Wildlife, Conflict or Co-existence? 1st ed. New York: Cambridge University Press; 2005. 497 p

[7] DiStefano E. Human-Wildlife Conflict worldwide: collection of case studies, analysis of management strategies and good practices. Food and Agricultural Organization of the United Nations (FAO), Sustainable Agriculture and Rural Development Initiative (SARDI), Rome, Italy [Internet]. 2005. Available from: FAO Corporate Document repository http://www.fao.org/documents

[8] Berkes F. Rethinking community-based conservation. Conservation Biology. 2004;**18**:621-630

[9] TCH S, Ehringhaus C, Campbell BM. Conservation and development in tropical forest landscapes: A time to face the trade-offs? Environmental Conservation. 2008;**34**:276-279

[10] Gigliotti LM, Decker DJ. Human dimensions in wildlife management education: pre-service opportunities and in-service needs. Wildlife Society Bulletin (1973–2006). 1992;**20**:8-14

[11] Lee PC, Priston NE. Human attitudes to primates: Perceptions of pests, conflict and consequences for primate conservation. Commensalism and Conflict: The Human-Primate Interface. 2005;**4**:1-23

[12] Pedersen AB, Jones KE, Nunn CL, Altizer S. Infectious diseases and extinction risk in wild mammals. Conservation Biology. 2007;**21**:1269-1279

[13] Hill CM. Conflict of interest between people and baboons: Crop raiding in Uganda. International Journal of Primatology. 2000;**21**:299-315

[14] Hill CM. Farmers' perspectives of conflict at the wildlife–agriculture boundary: Some lessons learned from African subsistence farmers. Human Dimensions of Wildlife. 2004;**9**: 279-286

[15] Riley EP. The human–macaque interface: Conservation implications of current and future overlap and conflict in Lore Lindu National Park, Sulawesi, Indonesia. American Anthropologist. 2007;**109**:473-484

[16] Hill CM, Webber AD. Perceptions of nonhuman primates in human–wildlife conflict scenarios. American Journal of Primatology. 2010;**72**:919-924

[17] Martin RD. Origins, diversity and relationships of lemurs. International Journal of Primatology. 2000;**21**:1021-1049

[18] IUCN. IUCN Red List of threatened species. Version 2015–4. [Internet]. 2015. http://www.iucnredlist.org [Accessed: 2017-07-20]

[19] Tattersall I. Understanding species-level primate diversity in Madagascar. Madagascar Conservation & Development. 2013;**8**:7-11

[20] Waeber PO, Wilmé L, Ramamonjisoa B, Garcia C, Rakotomalala D, et al. Dry forests in Madagascar: Neglected and under pressure. International Forestry Review. 2015;**17**:127-148

[21] Mittermeier RE, Louis EE Jr, Langrand O, Schwitzer C, Gauthier CA, et al. Lémuriens de Madagascar. In: Publications scientifiques du Muséum national d'Histoire naturelle. Paris, Conservation International; 2014

[22] Rakotoarivelo AR, Razafimanahaka JH, Rabesihanaka S, Jones JPG, Jenkins RKB. Laws and regulations on wildlife in Madagascar: Progress and future needs. Madagascar Conservation & Development. 2011;**6**:37-44

[23] Borrini-Feyerabend G, Dudley N. Les Aires Protégées à Madagascar: bâtir le système à partir de la base. World Commission on Protected Areas & International Union for Conservation of Nature. 2005

[24] Norris S. Madagascar defiant. AIBS Bulletin. 2006;**56**:960-965

[25] Golden CD. Bushmeat hunting and use in the Makira Forest, north-eastern Madagascar: A conservation and livelihoods issue. Oryx. 2009;**43**:386-392

[26] Barrett MA, Ratsimbazafy J. Luxury bushmeat trade threatens lemur conservation. Nature. 2009;**461**:470-470

[27] RKB J, Keane A, Rakotoarivelo AR, Rakotomboavonjy V, Randrianandrianina FH, et al. Analysis of patterns of bushmeat consumption reveals extensive exploitation of protected species in eastern Madagascar. PLoS One. 2011;**6**:e27570

[28] Reuter KE, Gilles H, Wills AR, Sewall BJ. Live capture and ownership of lemurs in Madagascar: Extent and conservation implications. Oryx. 2016;**50**:344-354

[29] Schwitzer C, Mittermeier RA, Johnson SE, Donati G, Irwin M, et al. Averting lemur extinctions amid Madagascar's political crisis. Science. 2014;**343**:842-843

[30] Geist HJ, Lambin EF. Proximate causes and underlying driving forces of tropical deforestation: Tropical forests are disappearing as the result of many pressures, both local and regional, acting in various combinations in different geographical locations. Bioscience. 2002;**52**:143-150

[31] Irwin MT, Johnson SE, Wright PC. The state of lemur conservation in south-eastern Madagascar: Population and habitat assessments for diurnal and cathemeral lemurs using surveys, satellite imagery and GIS. Oryx. 2005;**39**:204-218

[32] Estrada A, Garber PA, Rylands AB, Roos C, Fernandez-Duque E, et al. Impending extinction crisis of the world's primates: Why primates matter. Science Advances. 2017;**3**: e1600946

[33] Michalski F, Peres CA. Anthropogenic determinants of primate and carnivore local extinctions in a fragmented forest landscape of southern Amazonia. Biological Conservation. 2005;**124**:383-396

[34] Davies TJ, Pedersen AB. Phylogeny and geography predict pathogen community similarity in wild primates and humans. Proceedings of the Royal Society of London B: Biological Sciences. 2008;**275**:1695-1701

[35] Bublitz DC, Wright PC, Rasambainarivo FT, Arrigo-Nelson SJ, Bodager JR, Gillespie TR. Pathogenic enterobacteria in lemurs associated with anthropogenic disturbance. American Journal of Primatology. 2015;**77**:330-337

[36] Mittermeier RA, Tattersall I, Konstant WR, Meyers DM, Mast RB, Nash SD. Lemurs of Madagascar. Washington, DC, USA: Conservation International; 1994

[37] Golden C. Spiritual roots of the land. Worldviews: Global Religions, Culture, Ecology. 2014;**18**:255-268

[38] Reibelt LM, Richter T, Rendigs A, Mantilla-Contreras J. Malagasy conservationists and environmental educators: Life paths into conservation. Sustainability. 2017;**9**:227. DOI: 10.3390/su9020227

[39] Simons EL, Meyers DM. Folklore and beliefs about the Aye aye (*Daubentonia madagascariensis*). Lemur News. 2001;**6**:11-16

[40] Ganzhorn JU, Abraham JP. Possible role of plantations for lemur conservation in Madagascar: Food for folivorous species. Folia Primatologica. 1991;**56**:171-176

[41] Anonymous. News and views. Oryx. 1964;**7**:148

[42] Gezon LL. Political ecology and conflict in Ankarana, Madagascar. Ethnology. 1997;**36**:85-100

[43] Peters J. Understanding conflicts between people and parks at Ranomafana, Madagascar. Agriculture and Human Values. 1999;**16**:65-74

[44] Neumann RP. Imposing Wilderness: Struggles Over Livelihood and Nature Preservation in Africa. 4th ed. Berkeley and Los Angeles: University of California Press; 2001. 257 p

[45] Mehta JN, Kellert SR. Local attitudes toward community-based conservation policy and programmes in Nepal: A case study in the Makalu-Barun conservation area. Environmental Conservation. 1998;**25**:320-333

[46] Murphree MW. 2000. Community–based conservation: Old ways, new myths and enduring challenges. Conference on African wildlife Management in the new Millennium, 13–15 December 2000, Mweka, Tanzania. [Internet]. 2000. [Available: http://goo.gl/svnADC]

[47] Gezon LL. Who wins and who loses? Unpacking the "local people" concept in ecotourism: A longitudinal study of community equity in Ankarana, Madagascar. Journal of Sustainable Tourism. 2014;**22**:821-838

[48] Reibelt LM, Nowack J. Community-based conservation in Madagascar, the 'cure-all' solution? Madagascar Conservation & Development. 2015;**10**:3-5

[49] Rendigs A, Reibelt LM, Ralainasolo FB, Ratsimbazafy JH, Waeber PO. Ten years into the marshes–*Hapalemur alaotrensis* conservation, one step forward and two steps back? Madagascar Conservation & Development. 2015;**10**:13-20

[50] Waeber PO, Wilmé L, Mercier J-L, Rakotozafy LMA, Garcia C, Sorg J-P. The role of lakes in the context of the centers of endemism. Akon'ny Ala. 2015;**32**:34-47

[51] Waeber PO, Ralainasolo FB, Ratsimbazafy JH, Nievergelt CM. Consequences of Lakeside Living for the Diet and Social Ecology of the Alaotran Gentle Lemur. In: Primates in Flooded Habitats: Ecology and Conservation. Cambridge: Cambridge University Press; 2018 In press

[52] Waeber PO, Ratsimbazafy JH, Andrianandrasana H, Ralainasolo FB, Nievergelt CM. *Hapalemur alaotrensis*, a Conservation Case Study from the Swamps of Alaotra, Madagascar. In: Primates in Flooded Habitats: Ecology and Conservation. Cambridge: Cambridge University Press; 2018 In press

[53] Mutschler T, ATC F, Nievergelt CM. Preliminary field data on group size, diet and activity in the Alaotran gentle lemur *Hapalemur griseus alaotrensis*. Folia Primatologica. 1998;**69**:325-330

[54] Waeber PO, Hemelrijk CK. Female dominance and social structure in Alaotran gentle lemurs. Behaviour. 2003;**140**:1235-1246

[55] IUCN. Andriaholinirina N, Baden A, Blanco M, Chikhi L, et al. *Hapalemur alaotrensis*. The IUCN Red List of Threatened Species. [Internet]. 2014. DOI: 10.2305/iucn.uk.2014-1.rlts.t9676a16119362.en

[56] Reibelt LM, Woolaver L, Moser G, Randriamalala IH, Raveloarimalala LM, et al. Contact matters: Local people's perceptions of *Hapalemur alaotrensis* and implications for conservation. International Journal of Primatology. 2017;**38**:588-608

[57] Ratsimbazafy JR, Ralainasolo FB, Rendigs A, Mantilla-Contreras J, Andrianandrasana H, et al. Gone in a puff of smoke? *Hapalemur alaotrensis* at great risk of extinction. Lemur News. 2013;**17**:14-18

[58] Raveloarimalala LM, Reibelt LM. Update on the management of Park Bandro and population numbers of *Hapalemur alaotrensis*. Lemur News. 2017;**20**:2

[59] Rittel HW, Webber MM. Dilemmas in a general theory of planning. Policy Sciences. 1973;**4**:155-169

[60] Game ET, Meijaard E, Sheil D, McDonald-Madden E. Conservation in a wicked complex world; challenges and solutions. Conservation Letters. 2014;**7**:271-277

[61] Waeber PO, De Grave A, Wilmé L, Garcia CA. Play, learn, explore: grasping complexity through gaming and photography. Madagascar Conservation & Development. DOI: 10.4314/mcd.wetlands.1

[62] Lammers PL, Richter T, Waeber PO, Mantilla-Contreras J. Lake Alaotra wetlands: How long can Madagascar's most important rice and fish production region withstand the anthropogenic pressure? Madagascar Conservation & Development. 2015;**10**:116-127

[63] Stoudmann N, Waeber PO, Randriamalala IH, Garcia C. Perception of change: Narratives and strategies of farmers in Madagascar. Journal of Rural Studies. 2017;**56**:76-86

[64] Ravaka A, Ramamonjisoa BS, Ratsimba Rakoto H, Ratovoson ANA. Circuit court du marché des produits agricoles: pour une gestion efficace du paysage ouvert, cas du

bassin-versant de Maningory, Madagascar. Madagascar Conservation & Development. DOI: 10.4314/mcd.wetlands.2

[65] Waeber PO, Wilmé L. Madagascar rich and intransparent. Madagascar Conservation & Development. 2013;**8**:52-54

[66] Waeber PO, Reibelt LM, Randriamalala IH, Moser G, Raveloarimalala LM, et al. Local awareness and perceptions: Consequences for conservation of marsh habitat at Lake Alaotra for one of the world's rarest lemurs. Oryx. DOI: 10.1017/S0030605316001198

[67] Reibelt LM, Richter T, Waeber PO, Rakotoarimanana SHNH, Mantilla-Contreras J. Environmental education in its infancy at Lake Alaotra, Madagascar. Madagascar Conservation & Development. 2014;**9**:71-82

[68] Rakotoarisoa TF, Waeber PO, Richter T, Mantilla-Contreras J. Water hyacinth (*Eichhornia crassipes*), any opportunities for the Alaotra wetlands and livelihoods? Madagascar Conservation & Development. 2015;**10**:128-136

[69] Zosso C. Marshland Management in the Alaotra Region (Madagascar) – Discussing Preferences with Local Stakeholders on the Basis of a Role-Playing Game. ETH Zurich, Zurich, Switzerland: Unpubl. M.Sc. thesis; 2016

[70] Pidgeon M. An Ecological Survey of Lake Alaotra and Selected Wetlands of Central and Eastern Madagascar in Analyzing the Demise of Madagascar Pochard *Aythya innotata*. Antananarivo, Madagascar: WWF/Missouri Botanical Garden; 1996

[71] Razanadrakoto D. Rapport Annuel 2003 CIRPRH. Ambatondrazaka, Madagascar: Circonscription de la Pêche et des Resource Halieutique; 2004

[72] Andrianandrasana HT, Randriamahefasoa J, Durbin J, Lewis RE, Ratsimbazafy JH. Participatory ecological monitoring of the Alaotra wetlands in Madagascar. Biodiversity and Conservation. 2005;**14**:2757-2774

[73] APC W, Milner-Gulland EJ, JPG J, Bunnefeld N, et al. Quantifying the short-term costs of conservation interventions for fishers at Lake Alaotra, Madagascar. PLoS One. 2017;**10**: e0129440

[74] Ranarijaona HLT. Concept de modèle écologique pour la zone humide Alaotra. Madagascar Conservation & Development. 2007;**2**:35-42

[75] Bakoariniaina LN, Kusky T, Raharimahefa T. Disappearing Lake Alaotra: Monitoring catastrophic erosion, waterway silting, and land degradation hazards in Madagascar using Landsat imagery. Journal of African Earth Science. 2006;**44**:241-252

[76] Wright HT, Rakotoarisoa JA. Human ecology. In: Goodman SM, Benstead JP, editors. The Natural History of Madagascar. Chicago: The University of Chicago Press; 2003. pp. 112-178

[77] Sunderlin WD, Angelsen A, Belcher B, Burgers P, Nasi R, et al. Livelihoods, forests, and conservation in developing countries: An overview. World Development. 2005;**33**:1383-1402

[78] Bamford AJ, Razafindrajao F, Young RP, Hilton GM. Profound and pervasive degradation of Madagascar's freshwater wetlands and links with biodiversity. PLoS One. 2017;**12**:e0182673

[79] Waeber PO, Wilmé L, Mercier JR, Camara C, Lowry IIPP. How effective have thirty years of internationally driven conservation and development efforts been in Madagascar? PLoS One. 2016;**11**:e0161115

[80] Wallace APC, Jones JP, Milner-Gulland EJ, Wallace GE, et al. Drivers of the distribution of fisher effort at Lake Alaotra, Madagascar. Human Ecology. 2016;**44**:105-117

[81] Reibelt LM, Moser G, Dray A, Randriamalala IH, Chamagne J, et al. Tool development to understand rural resource users' land use and impacts on land type changes in Madagascar. Madagascar Conservation & Development. DOI: 10.4314/mcd.wetlands.3

[82] Richter T, Rendigs A, Maminirina CP. Conservation messages in speech bubbles–evaluation of an environmental education comic distributed in elementary schools in Madagascar. Sustainability. 2015;**7**:8855-8880

[83] Etienne M. editor. Companion Modelling. A Participatory Approach to Support Sustainable Development. Éditions Quæ, Versailles, FR. 2014

[84] Myerson RB. Game Theory. Cambridge, Massachusetts and London: Harvard University Press; 2013

[85] Lindgren M, Bandhold H. Scenario Planning, The Link between Future and Strategy. New York: Palgrave Macmillan; 2003

[86] Levitt SD, List JA. What do laboratory experiments measuring social preferences reveal about the real world? The Journal of Economic Perspectives. 2007;**21**:153-174

[87] Lang DJ, Wiek A, Bergmann M, Stauffacher M, Martens P, Moll P, et al. Transdisciplinary research in sustainability science: Practice, principles, and challenges. Sustainability Science. 2012;**7**:25-43

[88] Etienne M, Du Toit D, Pollard S. ARDI: A co-construction method for participatory modeling in natural resources management. Ecology and Society. 2011;**16**:1-14

[89] Crookall D. Serious games, debriefing, and simulation/gaming as a discipline. Simulation & Gaming. 2010;**41**:898-920

[90] Adams WM. Future Nature: A Vision for Conservation. London: Earthscan; 2003

[91] Scales IR. The future of biodiversity conservation and environmental management in Madagascar: Lessons from the past and challenges ahead. In: Scales IR, editor. Conservation and Environmental Management in Madagascar. London: Routledge; 2014. pp. 342-360

[92] Treves A, Wallace RB, White S. Participatory planning of interventions to mitigate human–wildlife conflicts. Conservation Biology. 2009;**23**:1577-1587

[93] Jachmann H. Monitoring law-enforcement performance in nine protected areas in Ghana. Biological Conservation. 2008;**141**:89-99

[94] Hilborn R, Arcese P, Borner M, Hando J, Hopcraft G, Loibooki M, et al. Effective enforcement in a conservation area. Science. 2006;**314**:1266-1266

[95] Tranquilli S, Abedi-Lartey M, Amsini F, Arranz L, Asamoah A, Babafemi O, et al. Lack of conservation effort rapidly increases African great ape extinction risk. Conservation Letters. 2011;**5**:48-55

[96] Pfeifer M, Burgess ND, Swetnam RD, Platts PJ, Willcock S, Marchant R. Protected areas: Mixed success in conserving East Africa's evergreen forests. PLoS One. 2012;**7**:e39337

[97] Rasolofoson RA, Ferraro PJ, Jenkins CN, Jones JP. Effectiveness of community forest management at reducing deforestation in Madagascar. Biological Conservation. 2015;**184**: 271-277

[98] Reibelt LM, Ratsimbazafy J, Waeber PO. Lac Alaotra bamboo lemur *Hapalemur alaotrensis* (Rumpler, 1975). In: Schwitzer C, Mittermeier RA, Rylands AB, Chiozza F, Williamson EA, Macfie EJ, Wallis J, Cotton A, editors. Primates in Peril: The World's 25 Most Endangered Primates 2016–2018. Arlington, VA: IUCN SSC Primate Specialist Group (PSG), International Primatological Society (IPS), Conservation International (CI), and Bristol Zoological Society. pp. 32-34

Chapter 5

The Time-Budget Perspective of the Role of Time Dimension in Modular Network Dynamics during Functions of the Brain

Daya S. Gupta and Silmar Teixeira

Additional information is available at the end of the chapter

http://dx.doi.org/10.5772/intechopen.70588

Abstract

Information processing plays a key role in the daily activities of human and nonhuman primates. Information processing in the brain, underlying behavior, is constrained by the four-dimensional nature of external physical surroundings. In contrast to three geometric dimensions, there are no known peripheral sensory organs for the perception of time dimension. However, the representation of time dimension in modular neural networks is critical for the brain functions that require interval timing or the temporal coupling of action with perception. Recent experimental and theoretical studies are shedding light on how the representation of time dimension in neural circuits plays a key role in the diverse functions of the brain, which also includes motor interactions with environment as well as social interactions, such as verbal and nonverbal communication. Although different lines of evidence strongly suggest that rhythmic neural activities represent time dimension in the brain, how the information represented by rhythmic activities is processed to time behavioral responses by the brain remains unclear. Theoretical considerations suggest that the rhythmic activities represent a physical aspect of the time dimension rather than the source of simple additive temporal units for coding time intervals in neural circuits.

Keywords: time dimension in the brain, mirror neuron, modular network in the brain, visuomotor synchronization, muscle synergy, interval timing, movement timing

1. Introduction

The manner in which a primate allocates its waking hours to various activities is considered to represent an important aspect of its ecologic adaptation [1]. Surveys and analyses of time budget have been applied successfully to many diverse fields, which include mass media contact, service sector, urban planning, consumer behavior, the sexual division of labor, the informal economy and household economics, social accounting, social indicators, quality of life, way of life, social structure, etc. [2]. In this chapter, we will apply the time-budget analyses to various brain functions at subsecond to seconds range to understand the behavior of human and nonhuman primates during an interaction with their external surrounding.

Emerging data suggest that online behavior—the active response to changing demands in a current task—can be understood in terms of the allocation of various networks to distinct phases of a working memory function of the brain. Arguably, the human behavior must conform to the constraints of space-time fabric of surroundings, which makes cognitive time management an essential human function. Moreover, the importance of time budget is underscored by a recent meta-analysis, which concludes that common networks support modulation of efforts during nontemporal cognitive and timing tasks [3]. Defects of allocation of networks during online behavior may contribute to the development of psychiatric illnesses, such as schizophrenia, which characteristically exhibits a pattern of disconnectivity of timing circuits [4].

Time budgeting, the time allocation of various computing resources, available from distributed circuits in the brain to dynamic networks underlying brain functions, plays an important role in the online behavior of primates at the level of both individuals and groups. We will argue how the allocation of the brain's limited computing resources plays a crucial role in achieving optimal interaction of brain circuits at the level of both individuals and groups.

2. Dynamic interaction of primates with external space-time fabric directs the resource allocation to brain networks

Interval timing functions of the brain have played a key role in the survival of the human species during most of their existence as humans have been sustained by hunting and foraging [5]. Various activities, essential for the survival of humans, required interaction with external physical surrounding, with a spectrum of space-time fabric, which include stationary bodies of water to cascading streams, plains to mountains, diverse to simple to complex fauna and flora, presenting complex challenges for the human brain. Meeting diversely complex challenges requires the temporal coupling of various functions of the brain as well as processing time intervals on scales that may vary from subsecond range to several seconds.

Physical surroundings with which primate brain interacts is four-dimensional, three-geometric dimensions and the time dimension. Psychologic time has been a subject of intellectual curiosity for most of the known history, but the time dimension was studied as the fourth physical dimension only recently [6, 7]. Time dimension, unlike other physical qualities, is never perceived as a novelty, but only reported as the flow of time [8], and therefore, it is not easy to study by observation alone.

During interactions with external surroundings, dynamic networks are formed by synchronizing the activities of neurons, which results in the temporal coupling of information by distributed local circuits in the network [9]. Local circuits in networks generate patterns of neuronal activities in multidimensional domains, which represent information encoding motor response, sensory perception, as well as timing the behavior. For example, when someone catches a ball in flight or a fruit falling from a tree, brain networks produce information encoding the following: a series of muscle activation patterns called muscle synchrony [10, 11], resulting in the catch, timing muscle contractions and the time estimation of the arrival of the object at a suitable height. Activities of neurons, forming networks, produce synchronous neuronal activity patterns that represent information, which is processed during a successful interaction of the organism with its environment.

As suggested by the above examples, a successful interaction of the primate brain with four-dimensional space-time fabric of the external physical world requires interval timing functions at various time scales. A distributed modular neural clock mechanism is proposed by Gupta [12] to explain timing functions of the brain during online behavior.

2.1. Modular connections between small cortical information-processing areas establish dynamic networks that form the basis of timing the behavior

The modular nature of connections of local circuits forming dynamic networks in the cortex can be understood based on cytoarchitectural and electrophysiologic data. The cortical surface is divided into tiny computational units of millimeter range size, called the canonical microcircuits [13]. The neurons forming the canonical microcircuits have limited but conserved patterns of inputs and outputs [14]. Although the neurons within a canonical microcircuit are interconnected in a specific manner, the connectivity rates between most neuron pairs in the cortex are very low, rising only to 10–20% in specific cases when they are co-tuned to the same stimulus [14]. Since there is a low level of connectivity among neurons, it gives the canonical microcircuits the flexibility to form multiple configurations of neuronal circuits, and therefore, the ability to perform a wide variety of computations. This feature is useful for the role of small areas of the cortex to serve as relatively independent modules in neural networks.

Moreover, when inputs going to the cortex, primarily from the thalamus, are relayed in a topographically specific manner, there is a very little overlap between the inputs received by canonical circuits due to the low level of connectivity in the horizontal direction. Thus, this results in sparsely interconnected small divisions of the cortex, which act as independent circuits or modules that can be connected by synchronization. Synchronization of local circuits is due to the oscillating states of excitability and inhibition, which allows neurons to fire during a specific phase of a long-range oscillation when neurons are excitable: coupling the modules of a neural network. Periodic excitability of neurons during synchronization, due to the pacing by inhibitory neurons, produces oscillating extracellular currents that are recorded as neural oscillations [15].

2.2. Distributed modular neural clock mechanisms are responsible for timing behavior in primates

Modular model of distributed neural clocks is proposed by Gupta [12] for interval timing functions of the brain, such as timed-motor movements, time reproduction and time estimation. As depicted in the schematic in **Figure 1**, the proposed neural clock mechanism has three main modular components [12]: (a) calibration module, which are sensory and motor circuits of the brain that are involved in feedback interaction with the external four-dimensional surrounding,

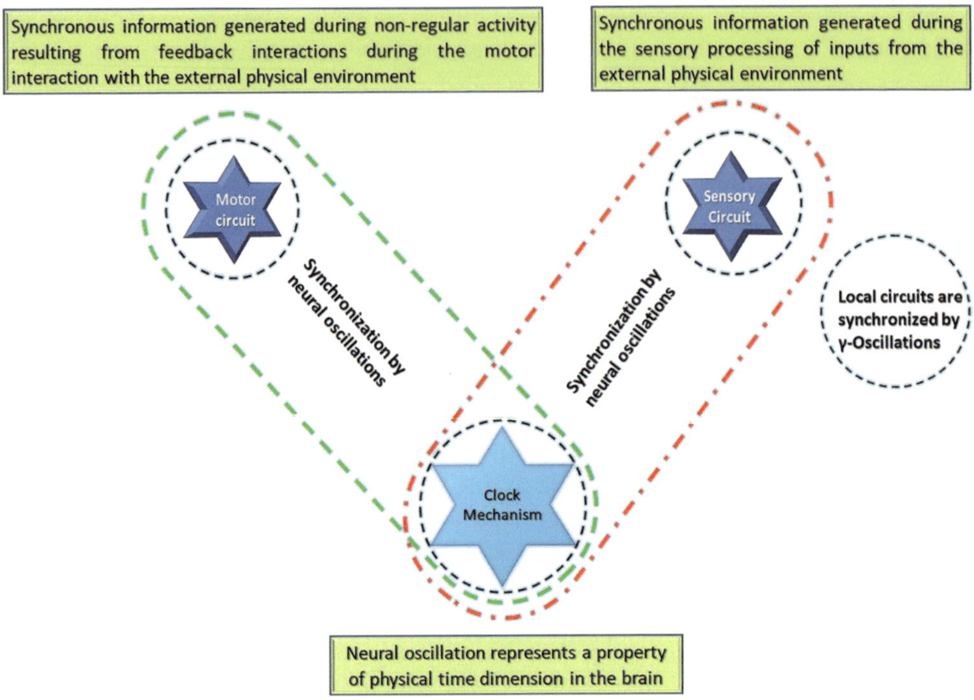

Figure 1. Schematic diagram depicts the modular clock model [12]:

- Synchronization (shown by dotted line of ovals) of neural clock mechanism with calibration modules: motor and sensory circuits, processing motor and sensory information during the interaction with space-time fabric of the external environment.
- Information is coded by the patterns of neural activities, which include spikes, spike bursts, logic states of circuits and slopes of ramping activity neurons.
- Synchronization is due to coherent, but oscillatory increase in the excitability of neurons in the network during specific phase of neural oscillations.
- Synchronization is responsible for the temporally coupling of information, produced by different neural activities, which encodes timed behavior.

(b) endogenous neural oscillator to represent physical time in neural circuits and (c) a clock mechanism for timing a behavioral response.

The functional role of the calibration module in the neural clock mechanism is to transfer information about physical time into neural circuits. Physical time information is transferred into neural circuits when motor and sensory information is processed by feedback mechanisms during an interaction of the brain with the external surroundings. During the feedback interaction, circuits associated with the motor and sensory functions produce neuronal activities that parallel the interactions between effector organs, muscles, sensory organs and the external space-time fabric. Moreover, the comparison of the intervals between changes outside the body and the intervals between corresponding feedback changes in neuronal activities in the brain would serve as a basis for the calibration of neural clocks.

An alternate mechanism for the calibration module is the cortico-basal ganglia-thalamocortical circuit, which, due to reciprocal connections and the consequent feedback process, could globally calibrate the neural circuits of the brain for the optimized interaction with external physical surroundings. This calibration mechanism predicts the presence of cortical neurons that are strongly coupled to population activity, but invariant to the stimulus conditions, which have been detected in monkey and mouse visual cortex [16].

Endogenous neural oscillator is the second component of the proposed modular neural clock mechanism. Neural oscillators are the rhythmic neural activities within the brain, such as neural oscillations, periodic bursts or rhythmic circuits. The idea of neural oscillator to represent time dimension is very old which is based on the intuitive role of the pendulum in mechanical clocks. Treisman [17] had originally proposed pacemaker-accumulator model. According to this model, a neural oscillator generates pulses, which are accumulated by a counter to encode time intervals in neural circuits.

Instead of serving as the source for temporal units for pulse accumulation, as in the Treisman model, the neural oscillation in the modular clock mechanism represents only a property of physical time. Thus, note that the periodicity of the endogenous oscillator does not simply represent a number that is added numerically to process time intervals for neural or psychologic processes. However, as mentioned later, the numerical quantification of time intervals in neural processes is likely encoded by spike patterns and their temporal relationship. Neural oscillators, representing physical time, along with calibration module and various task-specific circuits, synchronously generate information in networks, forming modular clock mechanism (**Figure 1**) to encode timed behavior by the brain.

Task-relevant neural clock is the third module, which is present in various parts of the brain, depending on the nature of the task. For example, the neural timers for visual time reproduction tasks in seconds range are present in the right dorsolateral prefrontal cortex [18–20].

At present, it is not clear how neural patterns, representing information, are coded and decoded to represent behavior, such as timing movements or time estimation. It is likely that a combination of different patterns such as spike patterns, logic states of circuits and ramping activity of neurons play various roles in coding and decoding information, leading to timed-behavioral responses [12].

Quantitative measurements, such as time intervals, are likely represented in neural circuits in numerical representations [21], such as spike patterns, which can be read as the binary numbers [8, 9]. Variable rather than fixed size of the time-bin that contain spikes or spike bursts in a neuronal activity will play a role in the representation of spike patterns as Shannon information [8]. Gallistel [21] has noted that studies suggest that the information about behaviorally relevant quantities such as timing behavior is not represented by the rates of spikes but rather by the intervals between their arrivals at synapses. Although the neurobiological basis of information processing, underlying the timing of behavior, remains far from clear [21], some consensus is present, such as neurons encode sensory information using a small number of active neurons at a particular time point [22]. This view is consistent with the cytoarchitectonic data that show a low level of connectivity among the neurons of the cortex.

2.3. Stochastic processes underlie the synchronization of circuits during motor movements

Motor movements play a key role in the interaction of primates with external physical world. But how the synchronization of different circuits leads to meaningful motor interactions with physical environment is not well understood. In a study of spatial visuomotor error during a speeded reaching movement, it was found that while subjects' objective distributions, task-related representation of external surroundings, are unimodal, their internal representations of the external task are typically mixtures of a small number of distributions [23]. This suggests that the central nervous system uses many possible neural circuits initially and synchronizes them in a limited number of combinations, which is the result of the constraints imposed by the external physical conditions. This is also consistent with the influential uncontrolled manifold theory, which postulates that control of limb movements gives priority to spatial shape of the movement over the trajectories of the individual joints [24, 25]. Current consensus favors that the movement planning by the human brain is best characterized in external, task-relevant coordinates—external surroundings—such as the direction of movement of the hand in reaching a target in external space [25].

Studies in primates have shown that the central nervous system uses flexible combinations of a limited number of muscle synergies—defined as a relative level of muscle contraction—to produce a variety of motor behaviors [10, 11]. Since muscle synergies may vary between individuals [10], it suggests the basis of the individual differences in the cytoarchitecture of the cortex that accompany possible differences in circuits underlying activation patterns. Muscle synergies likely result from the synchronization of a limited combination of circuits, reflecting a limited number of limb movements, conforming to the spatial coordinates of the target.

In a study of reaching movement task, joint angle variability peaked mid-way during the task [26]. Another study showed that reaching movements are characterized by high accuracy of end results [27], which indicates that the least variability is present after effector—the hand—reaches the target. Together, these observations suggest that individual circuits are synchronized more tightly at the end of reaching task when the hand reaches the target. Tighter synchronization of a given number of circuits will reduce the variability during the time-series activation of muscle synergies in the execution of motor tasks.

The mechanistic explanation, wherein a stochastic selection process chooses from an initially larger number of processing circuits to fine-tune final movements to reach a stationary external target, can be extended to the movements for catching moving objects. In case of moving targets, an accurate four-dimensional internal representation will be required for a successful task. This will require that the activity of multiple circuits, during various stages of movements, must be optimized to produce an efficient temporal coupling at the endpoint of reaching movements. This optimization process will also require time budgeting via synchronization to allocate different computing resources of the brain.

2.4. Brain networks undergo dynamic changes in connections during working memory function

Since working memory function of the brain is the ability to maintain and manipulate information over periods of seconds [28], it plays a key role during the online interaction with the space-time fabric of external surroundings. The ability to maintain and manipulate information in a working memory task is correlated with those patterns of neuronal activities that persist in the prefrontal cortex after the stimuli that elicited them no longer exist [28–30]. The limited amount of information, available for the manipulation by working memory [31, 32], suggests that there can be only a limited number of active networks present at a given time during a phase of working memory function. Thus, the limited number of connections of active networks must dynamically change during a working memory task to update the information processing to meet new demands.

Studies have shown that the prefrontal cortex and parietal cortex are the important regions for working memory functions [33, 34]. Furthermore, consistent with the dynamic changes in the networks for working memory function, reflecting the brain's interaction with four-dimensional surroundings, a past study reported that the functional connectivity patterns of fronto-parietal region, across various task states, shifted more than other networks in the brain [35]. Another study demonstrated that the functional connectivity of the frontoparietal networks is greater in tasks demanding greater cognitive control in normal and schizophrenic groups but showed an overall deficit in schizophrenia [36]. Given that the parietal cortical areas are the sites for processing of sensory information, such as visual, spatial and multisensory [37–39], the above study underscores the importance of the role of working memory networks in the cognitive control of the interaction with external surroundings.

2.5. Role of the thalamus in regulating the network connections in the brain during various arousal states

The thalamus plays a key role in regulating the brain oscillations that characterize various arousal states (**Figure 2**). A great variety of wave frequencies and patterns is controlled by the thalamus, which is attributed to the electrophysiologic properties and connectivity patterns of the cortical, thalamic reticular (RE), corticothalamic neurons (CT) and thalamocortical (TC) neurons [40]. The RE nucleus is a neuronal sheet made of GABAergic cells that envelops most of the surface of the thalamus [41, 42]. Unlike specific relay nuclei, neurons of the RE nucleus do not project directly to the cortex, but they receive collaterals from TC and CT neurons [42]. RE neurons provide both feedforward and feedback inhibition to excitatory TC relay neurons [41, 43] (**Figure 2**).

Multiple TC cells provide excitatory projections to each RE cells, and in turn, multiple RE cells contact each TC cell [43]. The sleep spindles result from strong reciprocal connections between excitatory TC neurons and inhibitory RE neurons [43]. When RE neuronal terminals become active, they inhibit TC neurons. The TC neurons exhibit a postinhibition burst of 7–15 Hz frequency called sleep spindles. This postinhibition burst of activity in TC neurons excites the inhibitory RE neurons, which consequently inhibits TC neurons via reciprocal connections. The inhibition of TC neurons, relaying sensory signals, prevents the cortex from receiving sensory information during sleep [44].

The slow wave (0.5–1 Hz), which is characteristic of NREM sleep, originates in the frontal regions and propagates in the anterior to posterior direction with a rapid speed of 1.2–7 m/s. Since the slow wave synchronizes the entire brain from the thalamus to the cortex, it prevents independent synchronization of smaller areas of the brain, which as a result interferes with the ability of the brain to form networks, necessary for the interaction with external environment.

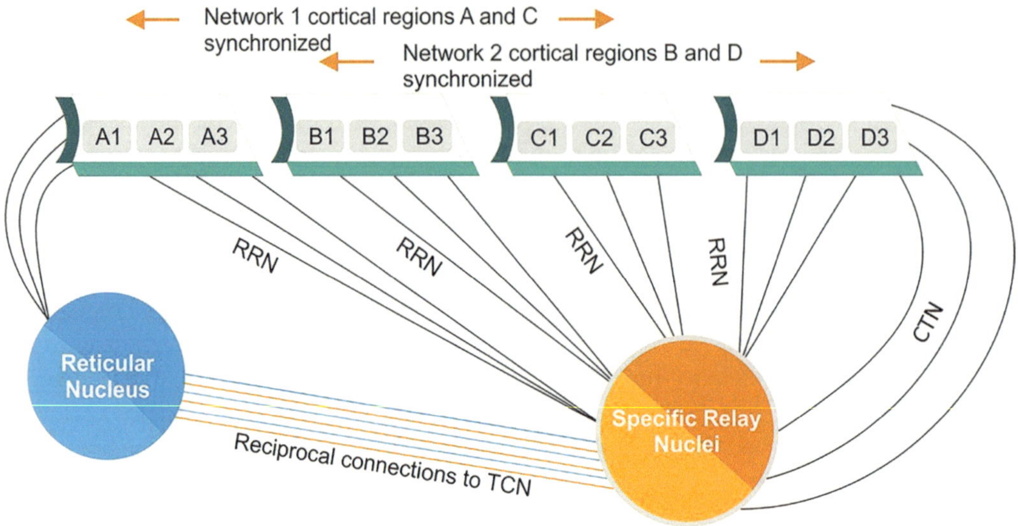

Figure 2. A schematic of connections between the cortex and thalamic reticular nucleus and specific thalamic nuclei:

- Cortical neurons in canonical microcircuits (shown as lettered gray rectangles, letters specify modules and numbers indicate component microcircuits in a module).
    - Canonical microcircuits receive inputs from rapid relay thalamocortical neurons (RRN).
    - Very scant horizontal connections are present between individual canonical microcircuits.
- There are rich divergent, convergent connections between the neurons of the reticular nucleus and thalamocortical neurons (TCN).
- There are heavy reciprocal and nonreciprocal projections from corticothalamic neurons (CTN) in layer 6 of the cortex.
- Hypothetical networks 1 and 2 result from the synchronization of different cortical areas representing modules.

However, the desynchronization of the slow wave paves the way for independent synchronization of smaller networks by neural oscillations during awaken state, covering distinct regions of the brain, making possible the online interaction with the environment. During an alert state, the slow wave oscillation is replaced with faster and short range and synchronized oscillations in the beta (15–30 Hz) and gamma (30–80 Hz) ranges [40, 45].

Note that the function of the thalamus is more than a passive relay center for sensory stimuli or the regulation of arousal states of the brain. Recent evidence indicates that the thalamus regulates functional connectivity between cortical microcircuits, which determines how cognitive processes are implemented [46]. TC neurons carry rapid sensory and motor relay information from specific thalamic nuclei project in a topographic fashion to sparsely interconnected microcircuits at the level of layer 4 of the cortex [14]. This enables the processing of relayed sensory information that occurs in modular neural circuits, comprising canonical microcircuit units.

Evidence suggests that the RE nucleus is an important hub in the communication between the thalamus and the cortex, which plays a key role in the cognitive processes that are affected in schizophrenia [47]. The RE nucleus covers the thalamus like an egg shell, and therefore, all fibers connecting the thalamus and the cortex must pass via the RE nucleus. Moreover, the anatomical data [48] argue against a global control of thalamocortical functions by the RE nucleus. Studies show that the RE nucleus can be subdivided into multiple sectors, each of which connects with a particular group of thalamic relay nuclei and cortical areas [48] and also due to a limited divergence from the thalamus to the cortex, the direct global control of cortical functions by the thalamus is highly improbable. Instead of global control, synchronization of small cortical areas can occur due to their direct specific connectivity to sectors in the RE nucleus. Moreover, it is plausible that the control of the function at the level of individual sectors of the RE nucleus can synchronize relatively distant areas to form dynamic networks in the cortex.

CT axons, exiting layer 6 [49], are more numerous than TC neurons synapsing in layer 4 [50]. CT axons provide massive inputs to the thalamus that are both reciprocal and nonreciprocal [50], and therefore can dynamically influence the excitability and sensory throughput of the thalamus [49]. Thus, the thalamus can help in the selection as well as the maintenance of dynamic networks during various task states of the brain. Moreover, the oscillations synchronizing network, due to the nonreciprocal connections of CT axons, can dynamically influence the periodicity of other oscillations, affecting timing functions by other circuits.

3. Timing circuits and neural networks during social interactions

3.1. Mirror neurons in sign language and vocalization: communication in humans and nonhuman primates

The ability to communicate, among individuals who are active in groups, has played a key role in the evolution of primitive human societies. Speech developed as a method of communication about 10 millenniums ago when creatures with large cranial capacity first appeared [5]. It is now believed that manual gestures are directly linked and have preceded the development of language in the humans as a method of communication [51]. Influential mirror system hypothesis posits that what counts for speaker (signer) must count approximately the same for hearer (observer) [52]. This parity rule underlies the basis of the development of communication in the nonhuman and human primates [51].

In an area of the macaque's ventral premotor cortex, area F5, which is homologous to the human Broca's area, the speech area, a new class of neurons called the mirror neurons, was discovered by Rizzolatti and his colleagues (**Figure 3**) [51, 53, 54]. The mirror neurons fire when monkeys perform a specific motor act or when they observe another primate, human or nonhuman, perform the same act [51, 53, 54]. More recently, Keysers, et al. [55] described a distinct population of neurons in the ventral premotor cortex of the monkey that discharges when the animal performs a specific action and it hears or sees the same action performed by an another individual. In contrast to the mirror neurons, these neurons, called audiovisual neurons, also fire when monkeys hear specific sounds related to an action. Studies of both classes of neurons have added considerably to our understanding of how primates read others' intention and produce context-related responses [56]. Series of studies of mirror neurons in F5 and the intraparietal sulcus, summarized in a review by Rozzi and Coude [56], have shown that depending upon the intention of motor act, for example grasping to eat or grasping to put into a container, the activities of neurons differed. Thus, the mirror neurons provide a mechanism to read motor intentions of other primates [56]. Imaging studies done in human subjects showed that the posterior parietal areas and premotor areas become active during action-observation and imitation [57, 58]. Rich reciprocal connections are present between different areas of the posterior parietal cortex and premotor cortex in monkeys, which serve as the anatomical basis of the mirror neuron mechanism (**Figure 3**) [56, 59]. The ability to read others' intentions would have played a key role during the human evolution due to the critical importance of this capability against the savage attacks by rival clans or species.

We note that the mirror and audiovisual neurons have an ability to synchronize parietal frontal circuits for the mirror system between two individuals. When two individuals are communicating by hand gestures or hear sounds representing the same action, then, the same neurons would fire in both the individuals: synchronizing the same set of circuits. This could be a basis of the development of communications among primates as the same sound or gesture may

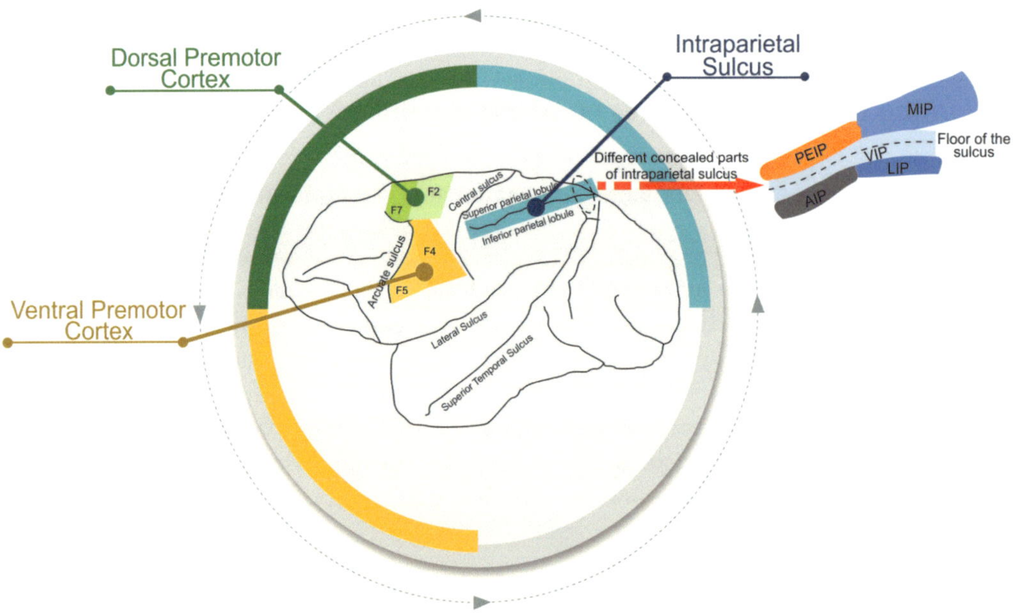

Figure 3. Anatomical basis of mirror neuron system in macaque brain:

- Ventral premotor cortex (areas represented by shades of yellow) contains frontal areas F5 and F4.
- Dorsal premotor cortex (areas represented by shades of green) contains frontal areas F7 and F2.
- Following parts of the intraparietal sulcus (blue) are depicted:
  - Anterior intraparietal area (AIP)
  - Medial intraparietal area (MIP)
  - Lateral intraparietal area (LIP)
  - Ventral intraparietal area (VIP).

lead to similar perception due to the activation of the same networks by common stimuli. The synchronization of common circuits by mirror neuron system would serve as a mechanism for same perception or meaning of a sound or gesture. It is believed that the Broca's area, the speech center of the human brain, was developed from the mirror neuron system in monkeys [51]. Moreover, F5 in the macaque brain, which contains mirror neurons and has a rich reciprocal connections with the posterior parietal cortex, is found to be homologous to the Broca's area based on the cytoarchitectonic and imaging data [60].

A recent study provides the evidence that the mirror neuron mechanism synchronizes networks spanning the ventral premotor cortex and posterior parietal cortex in the monkey brain (**Figure 3**) [61]. This study, which measured spiking activity, during a delayed grasping task by macaque monkeys, from the anterior intraparietal area (AIP) and ventral premotor area (**Figure 3**), both of which show strong connectivity [59], found that 18% of variability in reaction time was accounted by the ventral premotor area F5 but only 6% by the AIP. These observations suggest that there is a tighter coupling of mirror neurons in area F5 to physical time dimension, which is related to the greater contribution of the activity in F5 motor to the variability in reaction time. This study shows the presence of functional connectivity between the ventral motor area, F5, and the posterior parietal area, AIP, during the motor interaction with the external physical environment via delayed grasping task. This study is also consistent with the ability of mirror neuron system in F5 to synchronize the activity of the AIP in the posterior parietal cortex during a motor task.

When same mirror neurons in the brain of different individuals are stimulated by a common stimulus, then, the activity of those neurons will be locked to the same time point by the stimulus. This may serve as an important mechanism for the synchronization of perception of the shared space-time fabric of the surroundings among the individuals in a group. Without a mechanism to synchronize the perception by two individuals, who are communicating, a situation may develop analogous to two persons, who are watching the same action movie in two separate rooms, but the frames of movies being screened are fractions of a second apart. When they meet in a third room to discuss the last scene they had just watched, they will differ in their exact last version.

4. Representation of time dimension in the brain

4.1. Accurate representation of external physical time in the neural circuits is crucial for primate's interaction with the space-time fabric of external surroundings

Representation of time dimension in the brain is important for human and nonhuman primate's ability to survive. As we argued before, the representation of time dimension in neural circuits plays a key role in information processing underlying complex cognitive functions of the primate brain. Survival in many circumstances depends on the temporal coupling of actions with the perception of external environment.

Depending upon the demands of a task, the degree of coupling between action and perception may vary. For example, to dodge a falling rock requires a tighter temporal coupling between action and perception in comparison with the tasks to reach for a cup of water. To temporally couple actions with sensory inputs, an accurate representation of time information in neural circuits is required. Without an accurate representation of time dimension in neural circuits, the ability of the brain to successfully couple actions—motor movements dodging a falling rock—with the perception of falling rock would not be possible due to a mismatch with the external physical time scale. Tasks requiring a tighter coupling between action and perception, such as the ability to dodge a falling rock and to kill animals with projectiles, are likely to have played a significant role in the survival of humans and would have significantly guided the evolution of the human brain. Thus, it is interesting to note that the cerebellum evolved relatively more rapidly than the other parts of the human brain, including the neocortex [62].

4.2. Neural temporal unit as a measurement unit of time axis in the neural circuits

Representation of time units by regular events is inherent in the definition of regular events, which repeats itself after the same interval every time. Time units, such as seconds, measured by swings of pendulum in a mechanical clock, can help in measuring duration by counting the number of seconds or swings of a pendulum. Using this analogy, a neural temporal unit is defined as the interval between two adjacent regular spikes, spike bursts and is proposed to represent time units in neural circuits [12].

According to the pacemaker-accumulator model, when neural temporal units are added by the accumulator, it processes neural time intervals in a subjective or motor task. According to this model, if the neural temporal units represented by neural oscillators in the brain's timing circuits are smaller on physical-time scale, then subjective time reported in a task will be greater than the elapsed physical time. This will be the result of greater number of neural temporal units present within a given external time duration. As predicted by the pacemaker-accumulator model greater number of neural temporal units within a timed interval will lead to subjective over-estimation of intervals. This is supported by a study in which entrainment using visual flickers with faster frequency increased time measurement in a time-reproduction task [63]. Entrainment by faster flickers increases the frequency of neural oscillators in the brain, which leads to smaller temporal unit. Another study used auditory click trains to increase the speed of neural clocks, and studied its effects on pair-wise duration comparison and verbal time estimation task, and had arrived at similar conclusions [64]. However, not all entrainment studies agree with these conclusions [65]. Thus, a different role of neural oscillator is suggested within the modular clock model [12]. According to this formulation, the role of rhythmic activity is only to represent a physical property of the time dimension in neural clock mechanisms. Rhythmic activities are shown to be important for cognitive functions and various forms of behavior as reviewed by Herbst and Landau [66], but its precise role is yet to be understood.

4.3. Role of beta oscillations in the representation of time dimension

Accumulating body of evidence suggests that the beta-range neural oscillations represent physical time information in the brain (**Figure 1**) [65, 67–71]. A recent study has concluded that beta oscillations play an important role in the retention and manipulation of time information held in working memory [68]. In another study, where monkeys performed a synchronization-continuation task, after an initial increase in the beta power of local field potential recording in the striatum, there was a decrease in the beta power during the synchronization phase, which was followed by a rebound during the continuation phase [71]. The synchronization of the neural activity of the striatum by increase in beta oscillations in this study agrees with the networking of the components of cortico-basal ganglia-cortico-thalamic circuit in the neural clock mechanism during the continuation phase. Moreover, the causal relationship, between beta oscillations and the control of movements [72], also suggests that beta oscillations are responsible for coupling the neural-timer mechanism with the motor circuits for the control of movements.

4.4. Representation of time dimension in lower motor circuits

Central pattern generators (CPG) are networks of interneurons in the spinal cord forming a part of the hierarchical control by the central nervous system that plays a role in generating rhythmic motor activities in animals, such as walking and chewing [73]. The rhythmic activity of CPG networks, according to the formulations of distributed modular clock mechanism, represents the time dimension in spinal cord motor circuits that help maintain the temporal characteristics of locomotion. Although, CPG activity is observed after deafferentation or spinal cord injury, but sensory inputs, especially proprioceptive signals, are crucial for its role in locomotion [74]. The function of proprioceptive signals is chiefly the calibration of time dimension represented in rhythmic activity of the CPG during locomotion.

The evidence for the direct role of spinal cord CPG networks in human locomotion is scant and is mostly indirect [75, 76]. Some beneficial effects are seen in spinal cord injury patients following locomotor training [77], which can be explained by "learning by spinal cord CPG networks" about optimal incorporation of physical time-dimension information from sensory, especially proprioceptive inputs in lower motor circuits, processing movements after spinal cord injury.

5. Conclusion

There are many challenges that remain in trying to understand how the time dimension is incorporated in information processing that underlie the timing of behavior in primates. At present, we do not understand how time dimension represents information within the patterns of spikes or spike bursts. A key challenge is to understand how synchronous neural activity patterns in distinct local circuits in brain networks represent timed behavior during interaction with external world.

A better understanding of the answers to above questions will help us improve the management of a wide ranging group of illnesses, such as schizophrenia, Parkinson's disease and spinal cord injury among others.

Acknowledgements

We like to thank Marcos Ayres from the Brain Mapping and Cerebral Plasticity Laboratory, Federal University of Piauí, for the final figure artworks.

Author details

Daya S. Gupta[1]* and Silmar Teixeira[2]

*Address all correspondence to: dayagup@gmail.com

1 Department of Biology, Camden County College, Blackwood, NJ, USA

2 Brain Mapping and Plasticity Laboratory, Federal University of Piauí (UFPI), Parnaíba, Brazil

References

[1] Hemingway CA. Time budgets and foraging in a Malagasy primate: Do sex differences reflect reproductive condition and female dominance? Behavioral Ecology and Sociobiology. 1999;**45**(3):311-322

[2] Andorka R. Time budgets and their uses. Annual Review of Sociology. 1987;**13**:149-164

[3] Alustiza I et al. Meta-analysis of functional neuroimaging and cognitive control studies in schizophrenia: Preliminary elucidation of a core dysfunctional timing network. Frontiers in Psychology. 2016;**7**:192

[4] Ortuno F et al. Functional neural networks of time perception: Challenge and opportunity for schizophrenia research. Schizophrenia Research. 2011;**125**(2-3):129-135

[5] Maisels CK. The Emergence of Civilization: From Hunting and Gathering to Agriculture, Cities, and the State in the Near East. London, New York: Routledge; 1990. 395 p

[6] Neville EH. The Fourth Dimension. Cambridge Engineering: The University press; Cambridge, England. 1921. 4 p

[7] Petkov V, Minkowski H. Minkowski spacetime: A hundred years later. In: Fundamental Theories of Physics. Vol. 165. Dordrecht; New York: Springer; 2010 p. 1 online resource (xlii, 326 pages)

[8] Gupta DS, Merchant H. Editorial: Understanding the role of the time dimension in the brain information processing. Frontiers in Psychology. 2017;**8**:240

[9] Gupta DS, Chen L. Brain oscillations in perception, timing and action. Current Opinion in Behavioral Sciences. 2016;**8**:161-166

[10] Ting LH, McKay JL. Neuromechanics of muscle synergies for posture and movement. Current Opinion in Neurobiology. 2007;**17**(6):622-628

[11] Bizzi E et al. Combining modules for movement. Brain Research Reviews. 2008;**57**(1):125-133

[12] Gupta DS. Processing of sub- and supra-second intervals in the primate brain results from the calibration of neuronal oscillators via sensory, motor, and feedback processes. Frontiers in Psychology. 2014;**5**:816

[13] Miller KD. Canonical computations of cerebral cortex. Current Opinion in Neurobiology. 2016;**37**:75-84

[14] Harris KD, Shepherd GM. The neocortical circuit: Themes and variations. Nature Neuroscience. 2015;**18**(2):170-181

[15] Buzsaki G, Watson BO. Brain rhythms and neural syntax: Implications for efficient coding of cognitive content and neuropsychiatric disease. Dialogues in Clinical Neuroscience. 2012;**14**(4):345-367

[16] Okun M et al. Diverse coupling of neurons to populations in sensory cortex. Nature. 2015;**521**(7553):511-515

[17] Treisman M. Temporal discrimination and the indifference interval. Implications for a model of the "internal clock". Psychological Monographs. 1963;**77**(13):1-31

[18] Jones CR et al. The right dorsolateral prefrontal cortex is essential in time reproduction: An investigation with repetitive transcranial magnetic stimulation. Experimental Brain Research. 2004;**158**(3):366-372

[19] Koch G et al. Underestimation of time perception after repetitive transcranial magnetic stimulation. Neurology. 2003;**60**(11):1844-1846

[20] Ustun S, Kale EH, Cicek M. Neural networks for time perception and working memory. Frontiers in Human Neuroscience. 2017;**11**:83

[21] Gallistel CR. The coding question. Trends in Cognitive Sciences. 2017;**21**(7):498-508

[22] Olshausen BA, Field DJ. Sparse coding of sensory inputs. Current Opinion in Neurobiology. 2004;**14**(4):481-487

[23] Zhang H, Daw ND, Maloney LT. Human representation of visuo-motor uncertainty as mixtures of orthogonal basis distributions. Nature Neuroscience. 2015;**18**(8):1152-1158

[24] Bernshtein NA, The co-ordination and regulation of movements. Oxford, New York: Pergamon Press; 1967

[25] Latash ML et al. Motor control theories and their applications. Medicina (Kaunas, Lithuania). 2010;**46**(6):382-392

[26] Kruger M, Eggert T, Straube A. Joint angle variability in the time course of reaching movements. Clinical Neurophysiology. 2011;**122**(4):759-766

[27] Gordon J, Ghilardi MF, Ghez C. Accuracy of planar reaching movements. I. Independence of direction and extent variability. Experimental Brain Research. 1994;**99**(1):97-111

[28] Constantinidis C, Klingberg T. The neuroscience of working memory capacity and training. Nature Reviews Neuroscience. 2016;**17**(7):438-449

[29] Wallis JD, Anderson KC, Miller EK. Single neurons in prefrontal cortex encode abstract rules. Nature. 2001;**411**(6840):953-956

[30] Fuster JM, Alexander GE. Neuron activity related to short-term memory. Science. 1971;**173**(3997):652-654

[31] Cowan N. The magical number 4 in short-term memory: A reconsideration of mental storage capacity. The Behavioral and Brain Sciences. 2001;**24**(1):87-114 (discussion 114-85)

[32] Lewis RL. Interference in short-term memory: The magical number two (or three) in sentence processing. Journal of Psycholinguistic Research. 1996;**25**(1):93-115

[33] Eriksson J et al. Neurocognitive architecture of working memory. Neuron. 2015;**88**(1):33-46

[34] Vogel EK, Machizawa MG. Neural activity predicts individual differences in visual working memory capacity. Nature. 2004;**428**(6984):748-751

[35] Cole MW et al. Multi-task connectivity reveals flexible hubs for adaptive task control. Nature Neuroscience. 2013;**16**(9):1348-1355

[36] Ray KL et al. Functional network changes and cognitive control in schizophrenia. Neuroimage Clinical. 2017;**15**:161-170

[37] Bremmer F et al. Polymodal motion processing in posterior parietal and premotor cortex: A human fMRI study strongly implies equivalencies between humans and monkeys. Neuron. 2001;**29**(1):287-296

[38] Macaluso E, Driver J. Spatial attention and crossmodal interactions between vision and touch. Neuropsychologia. 2001;**39**(12):1304-1316

[39] Xing J, Andersen RA. Models of the posterior parietal cortex which perform multimodal integration and represent space in several coordinate frames. Journal of Cognitive Neuroscience. 2000;**12**(4):601-614

[40] Steriade M. Grouping of brain rhythms in corticothalamic systems. Neuroscience. 2006;**137**(4):1087-1106

[41] Pinault D. The thalamic reticular nucleus: Structure, function and concept. Brain Research. Brain Research Reviews. 2004;**46**(1):1-31

[42] Herrero MT, Barcia C, Navarro JM. Functional anatomy of thalamus and basal ganglia. Child's Nervous System. 2002;**18**(8):386-404

[43] Fogerson PM, Huguenard JR. Tapping the brakes: Cellular and synaptic mechanisms that regulate thalamic oscillations. Neuron. 2016;**92**(4):687-704

[44] Steriade M. Sleep oscillations and their blockage by activating systems. Journal of Psychiatry & Neuroscience. 1994;**19**(5):354-358

[45] Neske GT. The slow oscillation in cortical and thalamic networks: Mechanisms and functions. Front Neural Circuits. 2015;**9**:88

[46] Nakajima M, Halassa MM. Thalamic control of functional cortical connectivity. Current Opinion in Neurobiology. 2017;**44**:127-131

[47] Pratt JA, Morris BJ. The thalamic reticular nucleus: A functional hub for thalamocortical network dysfunction in schizophrenia and a target for drug discovery. Journal of Psychopharmacology. 2015;**29**(2):127-137

[48] Guillery RW, Feig SL, Lozsadi DA. Paying attention to the thalamic reticular nucleus. Trends in Neurosciences. 1998;**21**(1):28-32

[49] Crandall SR, Cruikshank SJ, Connors BW. A corticothalamic switch: Controlling the thalamus with dynamic synapses. Neuron. 2015;**86**(3):768-782

[50] Deschenes M, Veinante P, Zhang ZW. The organization of corticothalamic projections: Reciprocity versus parity. Brain Research. Brain Research Reviews. 1998;**28**(3):286-308

[51] Liebal K, Müller C, Pika S. Gestural Communication In Nonhuman and Human Primates. Benjamins Current Topics. Vol. xiv. Amsterdam, Philadelphia: John Benjamins Publishing Company; 2007. 284 p

[52] Rizzolatti G, Arbib MA. Language within our grasp. Trends in Neurosciences. 1998;**21**(5):188-194

[53] Gallese V et al. Action recognition in the premotor cortex. Brain. 1996;**119**(Pt 2):593-609

[54] Rizzolatti G et al. Premotor cortex and the recognition of motor actions. Brain Research. Cognitive Brain Research. 1996;**3**(2):131-141

[55] Keysers C et al. Audiovisual mirror neurons and action recognition. Experimental Brain Research. 2003;**153**(4):628-636

[56] Rozzi S, Coude G. Grasping actions and social interaction: Neural bases and anatomical circuitry in the monkey. Frontiers in Psychology. 2015;**6**:973

[57] Caspers S et al. ALE meta-analysis of action observation and imitation in the human brain. NeuroImage. 2010;**50**(3):1148-1167

[58] Molenberghs P, Cunnington R, Mattingley JB. Brain regions with mirror properties: A meta-analysis of 125 human fMRI studies. Neuroscience and Biobehavioral Reviews. 2012;**36**(1):341-349

[59] Grefkes C, Fink GR. The functional organization of the intraparietal sulcus in humans and monkeys. Journal of Anatomy. 2005;**207**(1):3-17

[60] Binkofski F, Buccino G. Motor functions of the Broca's region. Brain and Language. 2004;**89**(2):362-369

[61] Michaels JA et al. Predicting reaction time from the neural state space of the premotor and parietal grasping network. The Journal of Neuroscience. 2015;**35**(32):11415-11432

[62] Barton RA, Venditti C. Rapid evolution of the cerebellum in humans and other great apes. Current Biology. 2014;**24**(20):2440-2444

[63] Kanai R et al. Time dilation in dynamic visual display. Journal of Vision. 2006;**6**(12):1421-1430

[64] Penton-Voak IS et al. Speeding up an internal clock in humans? Effects of click trains on subjective duration. Journal of Experimental Psychology. Animal Behavior Processes. 1996;**22**(3):307-320

[65] Wiener M, Kanai R. Frequency tuning for temporal perception and prediction. Current Opinion in Behavioral Sciences. 2016;**8**:1-6

[66] Herbst SK, Landau AN. Rhythms for cognition: The case of temporal processing. Current Opinion in Behavioral Sciences. 2016;**8**:85-93

[67] Chang A, Bosnyak DJ, Trainor LJ. Unpredicted pitch modulates Beta oscillatory power during rhythmic entrainment to a tone sequence. Frontiers in Psychology. 2016;**7**:327

[68] Chen Y, Huang X. Modulation of alpha and Beta oscillations during an n-back task with varying temporal memory load. Frontiers in Psychology. 2015;**6**:2031

[69] Kononowicz TW, van Rijn H. Single trial beta oscillations index time estimation. Neuropsychologia. 2015;**75**:381-389

[70] Cirelli LK et al. Beat-induced fluctuations in auditory cortical beta-band activity: Using EEG to measure age-related changes. Frontiers in Psychology. 2014;**5**:742

[71] Bartolo R, Merchant H. Beta oscillations are linked to the initiation of sensory-cued movement sequences and the internal guidance of regular tapping in the monkey. The Journal of Neuroscience. 2015;**35**(11):4635-4640

[72] Feurra M et al. Frequency-dependent tuning of the human motor system induced by transcranial oscillatory potentials. The Journal of Neuroscience. 2011;**31**(34):12165-12170

[73] Haghpanah SA, Farahmand F, Zohoor H. Modular neuromuscular control of human locomotion by central pattern generator. Journal of Biomechanics. 2017;**53**:154-162

[74] MacKay-Lyons M. Central pattern generation of locomotion: A review of the evidence. Physical Therapy. 2002;**82**(1):69-83

[75] Molinari M. Plasticity properties of CPG circuits in humans: Impact on gait recovery. Brain Research Bulletin. 2009;**78**(1):22-25

[76] Iosa M et al. Editorial: Neuro-motor control and feed-forward models of locomotion in humans. Frontiers in Human Neuroscience. 2015;**9**:306

[77] Behrman AL, Harkema SJ. Locomotor training after human spinal cord injury: A series of case studies. Physical Therapy. 2000;**80**(7):688-700

Pictorial Competence in Primates: A Cognitive Correlate of Mirror Self-Recognition?

Parron Carole

Additional information is available at the end of the chapter

http://dx.doi.org/10.5772/intechopen.75568

Abstract

Alternative interpretation to the long-standing assertion that mirror self-recognition entails self-awareness suggests that mirror self-recognition rather refers to the ability to differentiate its own body from other objects of the environment. From this standpoint, individuals should be able to interpret the mirror reflection as a symbolic representation of the self and to map this image to an internal representation of self. The framework of this chapter is based on the assumption that the cognitive processing underlying self-recognition might be related to the capacity of processing mirror image as a symbolic representation of the real object. To support that purpose, the critical developmental and comparative literature on pictorial competence and self-recognition ability in human infants and primates are contrasted. Furthermore, relationship between mirror self-recognition and pictorial abilities are discussed based upon two experiments. We first observed the behavior of pictorially naive primates, with a realistic picture. We second assessed whether non-naïve chimpanzees, demonstrating or not self-recognition, would behave with a realistic picture. Finally, I propose a refined postulate that illustrates how the pictorial competence and self-recognition ability may co-develop. The intent of this model is to open up new perspectives for further explorations of self-recognition ability in primates.

Keywords: mirror self-recognition, picture processing, referential stimuli, developmental abilities, monkeys, great apes

1. Introduction

Once upon a time, in a castle, a queen was talking to her image in the mirror: "Mirror, mirror on the wall, who is the fairest one of all?" (based on "Snow White"; the Grimm's fairy tales, 1812). Rather curiously, Snow White's step-mother, needed the opinion of her magic mirror to get an idea about her own reflection, although she could see it by herself. This suggests that the true nature of someone's image in the mirror does not always lay in the eyes of the perceiver. The queen, at least, knew that the reflection in the mirror was hers. Is it the case for non-human primates (hereafter referred to as primates) when they face a mirror? If they are able to recognize themselves, is this have anything to do with self-awareness? Self-recognition implies "… that one can become the object of one's own attention…" and thus possesses the ability "…to infer correctly the identity of the reflection in the mirror as being one-self" [1]. In spite of extensive empirical investigations for the last five decades, this issue is still hotly debated in the field of comparative psychology. The first experimental account of mirror self-recognition (from now MSR) in primates was made by Gallup in the 1970s [2]. The development of an original methodology, that is, the "mark test", was based on prior observation that initial exposure of chimpanzees to a mirror first elicit social responses toward their reflection (e.g. threat or play), but after a while, this behavior eventually vanishes and switches to self-directed behaviors (i.e. inspection of some areas of the body that are only visible with a mirror) [3]. Although the author assumed that self-directed behaviors necessarily imply self-recognition, he attempted to get more objective measures of self-recognition by marking the face of the chimpanzees during anesthesia, in such a way that this mark could only be seen when the chimps looked at their reflection in the mirror. After recovery from anesthesia, some of the chimpanzees used the mirror as a tool to investigate the mark on their face, which was the first evidence that primates can recognize themselves in a mirror. From these results, Gallup concluded that MSR implies not only awareness, but "self-awareness" as these individuals "…are objects of their own attention and are aware of their own existence…" [4, 5]. According to the same author, if chimpanzees do possess the cognitive ability of self-awareness, they should also possess the ability to monitor one's own mental state and to impute knowledge or emotional states to other individuals. These cognitive abilities go far beyond self-recognition since they imply some skills that follow from theory of mind [6]. It must be acknowledged that this theoretical assumption is exciting as it supposes that primates master high-level social skills that were historically confined to humans such as the attribution of intent, deception, reciprocal altruism, empathy, reconciliation? However, it is still debated whether MSR involves these "high-level social and cognitive abilities" related to the theory of mind [7–9] or "lower level cognitive abilities", such as the capacity to differentiate the body self from other objects of the environment, and to attend to stimuli that come from that body self [10]. For disentangling between these two theories, one needs first to precisely define the cognitive capacities that underlie primates' understanding of the true nature of their image in the mirror.

The present chapter attempts to demonstrate that MSR requires at *a minimum* that mirror reflections are processed as external representations of the self, and not as the real self. Therefore, being able of self-recognition might be related to the capacity of processing picture as a symbolic representation of the real object [11], and not to demanding cognitive abilities such as "self-awareness". I will first review the main results provided by the literature on MSR and on picture processing abilities in primates. I will then describe two experiments, carried out with my collaborators, which were aimed at testing the hypothesis that MSR might be based upon the cognitive ability to comprehend symbols. In conclusion, I will discuss our findings in regard to the literature and subsequently propose a novel postulate on the co-developmental course of pictorial comprehension and MSR in primates.

2. Mirror self-recognition in primates: experimental considerations

The pioneering study by Gallup [2] has been at the origin of a now very large literature on MSR in primates. To sum up, that literature confirms that some primates species can use their reflection in a mirror for self-exploration (for a review, see [12]), but claims that MSR abilities strongly depends on the species under consideration. On the one hand, MSR has been demonstrated in the four species of anthropoid primates: chimpanzees, for example [13], bonobos, for example [14], orangutans, for example [15]. MSR in gorillas is much more debated. Some studies indeed showed that gorillas exhibit no mark directed touching [16] nor spontaneous mirror-guided self-exploration [17], while others suggest that gorillas are capable of self-recognition [18–21]. Still, results are controversial even with mirror procedure adapted to the behavioral gorillas' specificity, and aimed at enhancing the gorillas' exploration of the mirror reflection, for example [22–24]. Overall, even if results on gorillas' MSR ability are mixed, according to the studies reporting positive findings, the capacity for self-recognition is at least present in some individuals.

On the other hand, failures to convincingly demonstrate MSR in non-ape primate species appear recurrent in the literature (for a review, see [25]). When facing a mirror, most monkeys' species usually display social responses toward their reflection, treating the image as a conspecific (familiar or not) [26, 27]. Monkeys' social responses generally do not evolve toward self-directed responses [28, 29]. In order to enhance monkeys' self-directed behavior in front of the mirror, researchers attempted to diversify the experimental procedures. Spontaneous responses to a mirror versus photographs and real-time videos performed by Cotton Top-Tamarins were recorded [30]. Subjects showed more attentional responses and frequent non-aggressive looks when facing the mirror compared to the other conditions. However, they did not consistently generate self-oriented behaviors in front of the mirror. Capuchin monkeys' spontaneous behaviors were also recorded when they were facing live video images of themselves [31], but again monkeys did not show any sign of explicit self-recognition. An alternative experimental strategy consists in modifying the nature or the location of the classic dye mark. In an experiment [32] where the classic mark was replaced by an odorant chocolate paste, the marmoset monkeys, while facing a mirror, did not demonstrate any mirror-guided exploration. In another study [33], capuchin monkeys were trained to touch a mark directly visible on various body areas (forearms, calves, and abdomen) to reinforce their experience of the correspondence between their body and its image in the mirror. During the test phase, the mark was confined to the face but monkeys never used their reflection to touch it. Other authors [34, 35] tried to prompt self-recognition in capuchin monkeys by manipulating the mirror's characteristics (e.g. size and number of mirror, angled-mirror, etc.), but also failed to observe any MSR in their subjects.

In macaques, results of MSR ability are more controversial. Macellini and collaborators [36] have attempted to promote own body recognition in pig-tailed macaques, by locating the mark on their chest (only visible via a mirror). When facing the mirror, none of the subjects tried to touch the mark, thus suggesting that they did not relate the mirror reflection to their body. In another study, researchers recorded the spontaneous behavior of rhesus monkeys equipped with a head implant (which is a more salient mark) while the animals were facing a mirror [37]. Macaques performed self-directed behaviors (e.g. active exploration of the implant and of the genital areas) only in front of the mirror. The same monkeys, however, failed to show any sign of self-recognition in a conventional mark test. Altogether, this experimental procedure appears insufficiently controlled to definitely conclude that these macaques actually demonstrated self-recognition ability (for a comment, see [38]). In a recent challenging study [39], rhesus monkeys were trained to locate a visual-somatosensory stimulus projected on their head. Basically, an irritant laser-pointer light dot was directed at the subject's forehead, and could be both perceived on the skin as a hot spot and as a colored spot in a mirror. This explicit training regimen not only resulted in macaques passing of the classic mark test with dyes of various colors but also showed that macaques subsequently explored spontaneously some unseen body parts. Although this study raised appreciations, for example [40], and excited comments by the scientific community "Clever studies like the one of Chang et al. [39] help expose our preconceptions about ourselves and point the way toward deeper understanding of the way our brains, and the brains of other animals, construct reality and our place within it" [41], it also raised some criticisms. Indeed, according to Anderson and Gallup [42], the trained monkeys might not have really recognized their image in the mirror but might have simply learned to touch the unpleasant location on their face when seeing any monkey image. To tackle this critical issue, Chang and collaborators designed a second experiment [43], replacing the visual-somatosensory stimulus (an irritant spot) by a visual-proprioceptive training. Macaques were trained to locate a spot projected onto a surface in their close personal space, which was visible either directly or through the mirror reflection. In a first phase, macaques were trained to locate the spot when it was both directly visible and through the mirror. During the test phase, subjects failed to pass the mark test as they did not touch the spot when it was projected on themselves. However, in a second training phase, the spot was only visible via the mirror reflection. In the test phase, macaques passed the mark test. This last experiment strongly suggests that under appropriate experimental approach, monkeys are able to pass the mark test. It should encourage researchers to adapt training procedures in order to promote the monkeys' motivation to use the mirror reflection which could also enhance their motivation to explore their own faces.

3. Mirror self-recognition in primates: intra- and inter-species differences

3.1. The mark test: a proper index of self-recognition?

Massive inter-individual differences in MSR abilities have been repeatedly reported in primates [44]. Even in the chimpanzees, the most proficient species in this test, MSR is only accessible to a subset of individuals. Many of them fail the task, for example [45, 46]. One extreme explanation of these inter-individual discrepancies is that non self-recognizer subjects or species actually lack "self-awareness". Only some chimpanzees, orangutans, and gorillas have consequently a visual representation of their self [4, 47]. Yet, other factors may also play a crucial role in the noticeable inter-individual differences. First, among the great ape species, a clear bias exists toward chimpanzees in terms of numbers of animals tested in MSR. About 164 chimpanzees (estimation) have been tested, while the number of orangutans (estimated $N = 5$), bonobos ($N = 10$) and gorillas (estimated $N = 19$) tested is relatively small.

This actual bias may be at the origin of the common idea that chimpanzees are the most proficient species in MSR. Second, rearing conditions of animals are extremely heterogeneous (captive vs. wild-born, different laboratory living conditions) and may probably result either in enhanced social-cognitive functioning (e.g. in the case of some 'enculturated' home-reared apes) or in impaired social-cognitive functioning (e.g. in the case of social deprivation, or very small group size). Unfortunately, the precise rearing conditions and previous cognitive experience of many primates tested in MSR are often unknown or at least not sufficiently described. Researchers may have underestimate the cognitive consequences of social factors and previous experimental history in the MSR ability. Third, the standard version of the mark test is good only for positively proving the existence of self-recognition. Indeed, failures only sign an absence of self-directed behaviors but not a lack of self-recognition *per se* (false negatives) (for a review, see [48]). And finally, studies on MSR mostly look for chimpanzees-like behaviors in front of a mirror, as initially described by Gallup [2]. Species' behavioral repertoires may notably make it more difficult for an animal to detect the contingencies between its body and the mirror image, *a sine qua non* to adopt self-directed behaviors, and eventually achieve MSR. On the one hand, as primates initially mistake their own reflection in the mirror with a conspecific, the exploration of the mirror may substantially vary among species and thus provide different visual and kinaesthetic information, from a qualitative and quantitative point of view. Direct gazing in gorillas, for instance, may result in a response of fear from the subject looked at, or in gaze avoidance to sign appeasement and submission [49]. Macaques display threat gaze during conflicts and avert gaze during friendly approaches [50], while chimpanzees avert gaze in case of potential conflict and make eye contact to reconcile [51]. On the other hand, primates' spontaneous behaviors in front of a mirror may provide, depending on the innate characteristics of each species, more or less somato-sensitive information, which further promote self-directed behaviors. Monkeys, for instance, are usually less engage in auto-grooming than chimpanzees and thus do not receive direct feedback from their bodies as frequently as the chimpanzees do [52, 53]. Yet, the same mirror tests have been applied to a wide range of species, while it might be inappropriate to test some primates' species under these conditions given their own social characteristics.

3.2. Symbolic pictorial competence: an index of mirror self-recognition ability?

In the following, I will propose an alternative explanation to inter-species discrepancies in MSR. Before using the concept of "self-awareness", I would recommend first to analyze the perceptual information conveyed by the mirror image in order to infer the cognitive processes at work in the MSR task. When the monkeys look at a mirror, they often express social behaviors, for example [27]. This behavior suggests that they perceive a very realistic pictorial representation of a monkey, and probably mistake the mirror reflection of themselves with a real monkey: they process the "picture" as the "real object" itself. This perceptual phenomenon has already been described in the literature and named "confusion mode" of picture processing [54]. Apes express self-directed behaviors when facing a mirror, for example [13] thus suggesting that they do not perceive the mirror reflection as another individual but rather as a representation of themselves. According to Bard and collaborators [11], self-recognition implies "an understanding that the self can exist and can be represented: the mirror image is a representation of the self as an iconic symbol". This ability to understand the image as a symbolic representation infers an "equivalence mode" of picture processing [54]. This "equivalence mode" refers to a situation in which the subject associates the real object with its picture while being perfectly aware that the picture is different from the real object. In that sense, the picture, as well as the mirror image, is a referential stimulus: it is an object which represents another object. Animals must interpret the relation between the picture and its referent to use this medium as symbols and sources of information about the world. Self-recognition ability may thus appeal to the capacity of inferring the dual nature of the mirror reflection. An overview of the comparative literature on picture perception hereafter will highlight the conditions under which the pictures are processed, or not, as an iconic symbol by primates.

4. Picture perception by primates

Three levels of picture processing in monkeys can be outlined [54]: *Confusion, Independence and Equivalence*. The *confusion mode* refers to a situation in which the perceiver simply mistakes the real object and its depiction. The picture is processed as if it was the real object. The *independence mode* defines a situation in which the animal does not map the picture and the depicted object. The picture is processed based on its physical characteristics, regardless the meaning of the depicted object. The *equivalence mode* refers to a situation in which the animal processes the picture as a symbolic representation of the real object. The perceiver associates the real object with its picture while being perfectly aware that the picture is different from the real object. This equivalence mode is precisely the one, which might be required to interpret the reflection in the mirror as a representation of self.

4.1. The confusion mode of picture processing

Monkeys often react to pictures the way they would normally do in front of conspecifics: they express emotions. Macaques exhibit emotional responses (e.g. lip-smacking), when they observe pictures of faces [55], and express fear gestures in front of pictures of highly emotional objects [56]. Young macaques present signs of disturbance and vocalizations when seeing threat pictures [57]. In addition to pictorial stimuli conveying some emotions, pictures of highly motivating stimuli, that is, realistic food, has also been used to observe monkeys' reactions. The few studies existing showed that monkeys confused picture of food with real food and tended to process it, as they would normally do with real food. Bovet and colleagues were the first to illustrate this confusion behavior in [58]. Baboons were first trained to categorize food objects and non-food objects and showed positive transfer to novel objects. They were then trained with cut-out pictures of both food and non-food objects, and with the same pictures on a paper background. Categorical transfer occurred for cut-out photographs but not consistently for the whole ones, suggesting that the more pictures appear realistic, the more monkeys are deceived. Parron and collaborators [59] designed a simpler procedure in which pictorially naïve baboons had just to select one food picture (banana's picture) over a non-food one (pebble's picture). Results showed that subjects massively selected the realistic picture of banana and even, in some cases, ate this picture. Altogether, because these behaviors are perfectly appropriate in response to the presentation of real objects (i.e. a conspecific or food) but inappropriate in front of pictures, this suggests that monkeys recognized the depicted objects but did not process the pictures as some representations. To sum up, studies in monkeys suggest that naïve subjects processed realistic pictures in a confusion mode.

4.2. The independence mode of picture processing

A close analysis of the results reported in the literature leads to the conclusion that in many cases primates apply perceptual alternative strategies to perform picture recognition tasks. Under some specific experimental conditions, the animals process pictures based on a combination of perceptual features and patterns regardless of their representational content. For categorizing human versus non-human pictures, for example, capuchin monkeys used the colored features of the pictures, as humans' photos incidentally contained more red pixels than the non-human set of photos [60]. Two studies on face perception by baboons also provide an illustration of the perceptual strategies used by these animals to recognize pictures. Baboons were able to discriminate pictures of human faces on the sole basis of variations in the facial contour [61], and used the pixel similarities between training and probe pictures to categorize human versus baboon faces [62]. Apes, like monkeys, do not always process pictures as referential stimuli. For example, when tested in a matching to sample task using pictures, chimpanzees fail to show any transfer between real objects learnt by tactile inspection and pictures of the same objects [63]. The authors concluded that the "chimpanzees did not realize that a photograph is a visual stimulus that must be read". In another study, [64], where apes were tested to see whether they would be able to use pictures to infer the location of a hidden reward, results showed that iconic cue might help them to search for the reward but their poor performances eventually suggested that they did not fully grasp the informative content of the cue.

4.3. The equivalence mode of picture processing

Given the amount of studies on picture perception, we could expect to find more clear-cut evidence of a referential use of pictures by primates. For example, findings from a study on capuchin monkeys suggested allegedly positive evidence of an equivalence mode of picture processing [65]. After training, the subjects were able to discriminate images of in-group and out-group conspecifics' faces, and probably based their responses on the experience they have established with the depicted individuals in their real life. Nevertheless, as it will be detailed in Section 5, the experience animals had previously developed with tasks involving picture processing may have change drastically the way they interpreted the pictorial stimulus in this experiment. These capuchin monkeys received actually prior exposure to pictures of faces in other experiments, for example [66], which may have facilitated the positive transfer to grayscale images, and may also explained the absence of monkeys' social responses to pictures of faces. Comparable results were reported in crested macaques [67]. The subjects were able to discriminate members of their own social group from unfamiliar individuals, and were even better at recognizing higher ranking familiar individuals. As in [65], authors concluded that macaques applied their knowledge of their dominance hierarchies to the pictorial representation of their real group mates. Although this result rules out the independence mode of processing, the small sample size (three individuals who do not always perform similarly) along the absence of information on the past experience of these animals with pictures preclude unambiguous conclusions on picture perception in these monkeys. In [68], capuchin monkeys were able to match objects with their color photographs (and vice-versa), and even associate real objects with the corresponding black and white photographs, silhouettes and line drawings. However, the objects used in this experiment had no explicit representational content for the monkeys (e.g. a small wooden puppet or a cigarette lighter), and were never encountered in capuchins' real life before.

The most convincing evidence of an equivalence mode of picture processing comes from the so-called ape-language studies. Rumbaugh and collaborators [69] were the instigators of a pioneering project, "the Lana Project", at the Yerkes Regional Primate Research Center in Atlanta. This project was aimed at studying the language-like skills of apes, by using a new device designed to enhance teaching an ape to communicate with humans. So far, behavioral experiments studying language skills in apes suffered from a lack of experimental controlled conditions, slowing down the comprehension of apes' system of communication. Therefore, they developed an automated computer-controlled system: the chimpanzee was facing a keyboard displaying distinctive lexigrams and was trained to associate a real object or a picture to a specific symbol. This system actually improved both efficiency (e.g. increased number of trials per day) and objectivity (e.g. no interaction with a human experimenter and automated responses' recording).

This pioneering system was also used to study the ape-language skills by Savage-Rumbaugh and collaborators [70], at the Language Research Center in Atlanta. This project conducted to the historical first non-humans communication with humans by using symbolic language. Basically, some chimpanzees received, since a young age, extensive language training and first learned to use some lexigrams to name objects, and then to combine these lexigrams to produce a complex form of communication. With regard to our topic on pictorial competence in primates, we will consider the case of Sherman and Austin, two trained male chimpanzees, who were able to associate real food and tool objects to two lexigrams that served to "label" these categories [71]. Their performance in this "naming task" with real objects effectively transferred to pictorial depictions of tool and food items. They were able to make categorical judgments about objects when presented with only symbolic information, in the absence of the referent. They obviously did not confuse pictorial depictions with real objects. These two trained-language chimpanzees more likely processed the pictures considering their representational content, and not their pictorial expression.

Along the same line of research, Matsuzawa ([72] for an historic review), directly inspired by preceding ape-language studies conducted in Atlanta, developed a new project: "the Ai project" in 1978 at the Primate Research Institute in Kyoto University. Ai, a female chimpanzee, was intensively trained in various perceptual and cognitive tasks since her first year of age, with a computer-controlled apparatus. Over the years, she acquired a multitude of remarkable cognitive skills, including the use of visual symbols. Similarly to Sherman and Austin, she was trained to "name" some individuals by using lexigrams. When the same individuals were then represented on pictures [73], and line drawings [74], she could properly name them. Again, this result strongly suggests that chimpanzees may process the pictures considering their representational content, that is, in an equivalence mode. In another experiment, Tanaka [75] tested the ability of Ai, three other adult naive chimpanzees and three naive juveniles in a picture recognition task. The subjects were first trained to select some pictures of flowers among pictures of other natural objects. They were then tested with new pictures of flowers and some increasingly degraded stimuli: colored sketches, color clip art (cartoon-like pictures) and black and white line drawings of flowers. Ai was the unique adult chimpanzee to transfer from the flower picture to the non-photographic images of the flower. Noticeably, Ai is not only a language-trained chimpanzee, but she has also been previously exposed to line drawings [74]. The three juveniles, unlike the naïve adults, showed positive transfer of their previous category learning to the degraded stimuli. Their good performance may be due to their better learning ability, as it is established that young animals show greater cognitive flexibility, for example [76, 77]. Close and Call [78] used a similar procedure with sub-adult and adult naïve chimpanzees. The subjects were presented with colored, black and white sketches, and line drawings of the training pictures (tree vs. flower). Adolescent chimpanzees generally outperformed the adults, particularly in the recognition of the black and white sketches stimuli. None of the chimpanzees initially showed a positive transfer to the line drawings stimuli. However, the adolescent chimpanzees finally learned to categorize line drawings after a moderate period of training. These findings are broadly consistent with those of Tanaka [75], as after training, juvenile chimpanzees, unlike naïve adults, similarly showed in [75] and in [78] positive transfer of their previous category learning (colored pictures) to the degraded stimuli (black and white pictures or line drawings). Overall, it suggests that learning abilities may be enhanced by training during a critical period of chimpanzees' development that further enables adults to achieve various cognitive tasks.

5. The dynamic hypothesis of picture processing

The heterogeneity of the findings reported above suggests that some complex factors concurrently shape the perceptual and cognitive processes, and all together, this calls for a revision of the theoretical assumptions underlying picture perception in primates. Ontogeny, phylogeny, as well as the animal's level of experience with pictures seem to account for the observed performance variability in primate picture understanding. Actually, ***training regimen*** with pictorial display seems to influence drastically primates' behavioral responses. Untrained chimpanzees either failed to match real objects with their pictures when the objects were previously explored haptically [63] or to match real objects with line drawings of objects [75, 78]. It is only after training with degraded pictures that an equivalence mode of picture processing gradually emerged, at least in juveniles in these studies [75, 78]. A similar experiment with macaques trained to match objects and their pictures, and only when they were visible *at the same time,* also resulted in successful objects/pictures matching [79, 80].

The attractiveness of meaningful pictures (i.e. representing some objects of interest for monkeys) can be explained by the lack of experience of monkeys with this kind of new display, but may also vanish as a consequence of ***repeated exposures,*** which eventually result in habituation (i.e. decrease of response). When macaques are repeatedly presented with a "meaningless" picture (representing an abstract stimulus) and a "meaningful" picture (representing a real object), their response rate rapidly dropped for the "meaningful" picture, but remained constant for the "meaningless" picture [81]. In a two-alternative forced choice task, pictorially naïve baboons got rapidly habituated to pictures of -initially very attractive- realistic picture of bananas [59]. During the first test trial, 100% of the individuals participated but only half of the group responded three test trials later. In a recognition task using familiar versus unfamiliar faces, macaques showed a significant decrease in accuracy after 44 repetitions of the same stimuli, which suggests that they habituated to the stimuli and lacked interest in the task [67].

Noticeably, over the last decade, the development of experimental settings using touch screens to display pictures has significantly increased. Unlike procedure using real pictures, which not always allows animals to manipulate pictures, the use of touch screens may promote animal's understanding to the physical nature of pictures (e.g. flatness and unreachability of the depicted objects). This repeated experience probably leads animals to rapidly switch from a confusion mode of processing to an independence or an equivalence mode of processing, depending on the task demand and on the animals' cognitive abilities.

The past experience of primates with tasks involving picture understanding lead the same animals to a certain ***degree of expertise***. It should be reminded that the most proficient individuals in processing pictures as symbolic representations are the language-trained individuals, for example [71–74]. Indeed, Savage-Rumbaugh [70] evidenced that the learning process of symbols in apes is achieved by association. Indeed, they first needed an explicit training before they could use symbols referentially.

Based upon the above results of the literature, the animals' mode of picture processing seems not to be predetermined by the species but more likely by the subjects' own experimental history. Picture processing by primates is likely not limited to only one of the three different modes: confusion, independence, and equivalence. These modes evolved and should be regarded as components of a dynamic system, in which the trajectories between the three modes are under the influence of the animal's experience with pictures [82]. On the ground that pictorially naïve monkeys, just like young human infants [83], confuse realistic pictures with their referent objects [59], one can hypothesize that this confusion mode takes place at an early stage in the system of picture processing, likely the starting point in the dynamic scheme. From here, the experience with two-dimensional representation could direct the animals from the "confusion" starting point toward one of the two possible trajectories of the picture processing dynamic system. In the low-level road, experience (for instance, in the case of repeated

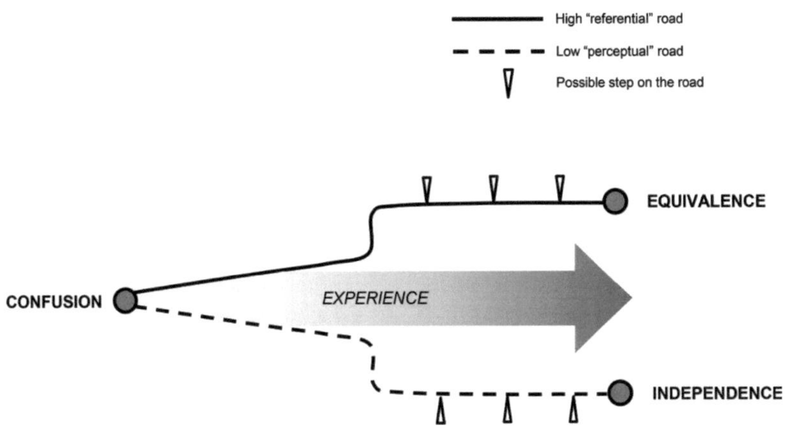

Figure 1. The three different levels of picture processing: confusion, independence, and equivalence and the two possible trajectories following experience with pictorial representations: the high "referential" road and the low "perceptual" road.

exposure and habituation) may push the animals toward the independence mode where the picture is defined by its physical characteristics and becomes meaningless. Another option is that experience, as in young human infants [84], drives the animals toward a high-level road or the road of equivalence mode, that is, the animals would ultimately be able to process pictures as iconic symbols (see **Figure 1**) [82]. According to the different cognitive processing underlying these two roads, from now, they will be referred as the low "perceptual" road and the high "referential" road. The capacity to take or not the high "referential" road correlates with other referential skills such as MSR will be assessed in the two studies detailed hereafter.

6. Empirical approach

In our first study, we showed that when pictorially naive baboons and gorillas looked at an attractive-realistic picture of banana for the first time, they often mistake the picture and the real object depicted. After smelling, or attempting to spell the depicted banana, they may even eat the picture. This striking behavior was only observed in baboons and gorillas, but never in chimpanzees. We thus hypothesized that the pictorial understanding of chimpanzees relying on an equivalence mode of processing might be at the origin of their MSR ability. They may process the mirror reflection as an iconic symbol and interpret it as a representation of their self. By contrast, the confusion mode of processing displayed by gorillas and baboons is consistent with their inability to pass the mark test. In order to assess this hypothesis, a second study comparing the chimpanzees' ability for MSR and picture understanding was running. We were expecting that the positive MSR chimpanzees would never mistake the banana picture with a real banana as they were also able to process their image in the mirror as a representation. By contrast, since they do not possess a referential ability, the non-self-recognizer chimpanzees would potentially process the banana picture like baboons and gorillas in our previous experiment.

6.1. Picture processing in monkeys and great apes

The first study reported here assessed the behavioral responses to photographs by pictorially naive baboons, gorillas and chimpanzees and was published in [59]. It involved 55 baboons (*Papio anubis*): 26 males and 29 females (mean age 6.9 years, S.D. = 4.6 years) from the CNRS Rousset-sur-Arc Primate Center (France), 7 chimpanzees (*Pan troglodytes*): 3 males and 4 females (mean age 9.75 years, S.D. = 7.3 years) from the Wolfgang Kohler Primate Research Center of the Leipzig Zoo, and 4 gorillas (*Gorilla gorilla*): 1 male and 3 females (age 24.5 years, S.D. = 11.15 years) from the zoos of Leipzig or Nuremberg (Germany). Monkeys and apes, habitually housed in social group, were either tested in groups or isolated, depending on the specificity of their living quarters. Their participation was always voluntary. All animals were selected at purpose because they have no known exposure to videos, still pictures, or drawings, and were thus pictorially naive. We hypothesized that testing monkeys and apes at an equivalent level of expertise with two-dimensional stimuli would provide cues on the evolution of picture comprehension within the primate phylum. This research adhered to the legal requirements of the current French laws and the European directive 2010/63/EU (see **Figure 2**).

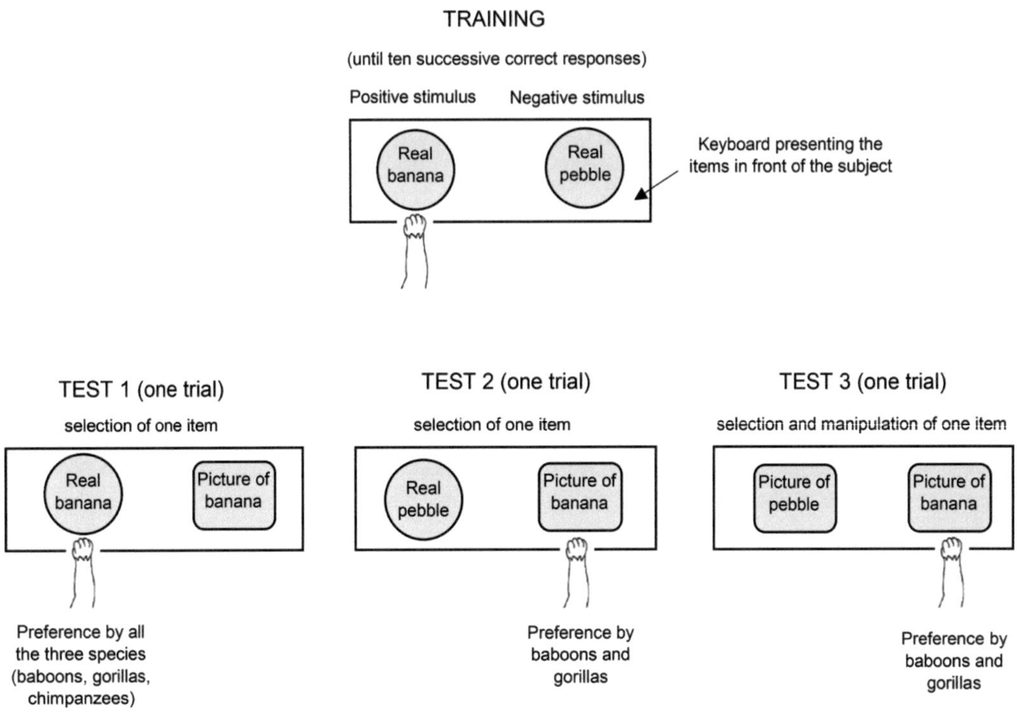

Figure 2. Illustration of the experimental protocol and summary of the results obtained for the three consecutive test trials. These results show that baboons, gorillas, and chimpanzees demonstrated a massive preference for the real banana compared to the picture of banana in test trial 1. When the subjects had to choose between the picture of banana and any other items in test trial 2 (the real pebble) and in test trial 3 (picture of pebble), baboons and gorillas clearly exhibited a preference for the picture of banana (*two-tailed binomial tests, $p < 0.05$ in all the three test conditions). Because of the small sample of chimpanzees, results preclude from any conclusion about the picture's preference in this naïve chimpanzees' group.

6.1.1. Procedure

A two-alternative forced choice task was used to train the subjects to select a real slice of banana (positive stimulus) over a real pebble (negative stimulus) when these items were presented simultaneously on two display panels. Subjects received a food reward when they selected the positive stimulus. These trials were repeated until a clear preference for the real banana slice emerged.

The test procedure consisted in three successive test trials per participant, using different pictures for each test that always subtended the same size as the real objects. Test trial 1 assessed the discrimination of a real slice of banana from a picture of a banana slice. Test trial 2 assessed the attractiveness of the three-dimensional real object in comparison to two-dimensional picture, by presenting a picture of a banana slice and a real pebble. Test trial 3 assessed whether preference for the real banana slice learned during training would transfer to the picture of the banana slice when the two real training objects (banana and pebble) were now presented as pictures. The participants could freely reached and manipulated the selected stimulus only in that last test trial.

6.1.1.1. Coding scheme

We coded the subjects' spontaneous behavior expressed with the picture during test trial 3; it was the very first manipulation of a picture, revealing the initial reaction of the naïve subjects with such an object. We used a hierarchical coding scheme to characterize the different behaviors observed during test 3 and grouped them in four categories. "touch or grasp"; "smell"; "bring to mouth"; "eat": in that last case, the animal ate the stimulus either partially or entirely. Chewing is observed and the animal ingests at least some parts of the stimulus (see images **Figure 3**).

6.1.2. Results

In test 1, all the species showed a clear preference for the real banana (vs. the picture of banana). In test 2, only baboons and gorillas showed a preference for the picture of banana (vs. the real pebble). In test 3, baboons and gorillas still showed a significant preference for the picture of banana (vs. the picture of pebble). In contrast, chimpanzees showed no clear preference for either picture. **Figure 4** shows the frequencies of each action obtained in baboons, gorillas and chimpanzees during the first bout of picture manipulation. Strikingly, 17 baboons and the 4 gorillas ate the picture of banana as if it was a real piece of banana. Some individuals even attempted to spell the depicted banana. These animals clearly did not process the picture of the banana as a representation. Interestingly, not only the chimpanzees expressed no reliable preference for the banana picture in these tests, but also never ate that stimulus, leaving open the possibility that the pictorial banana was a referential stimulus.

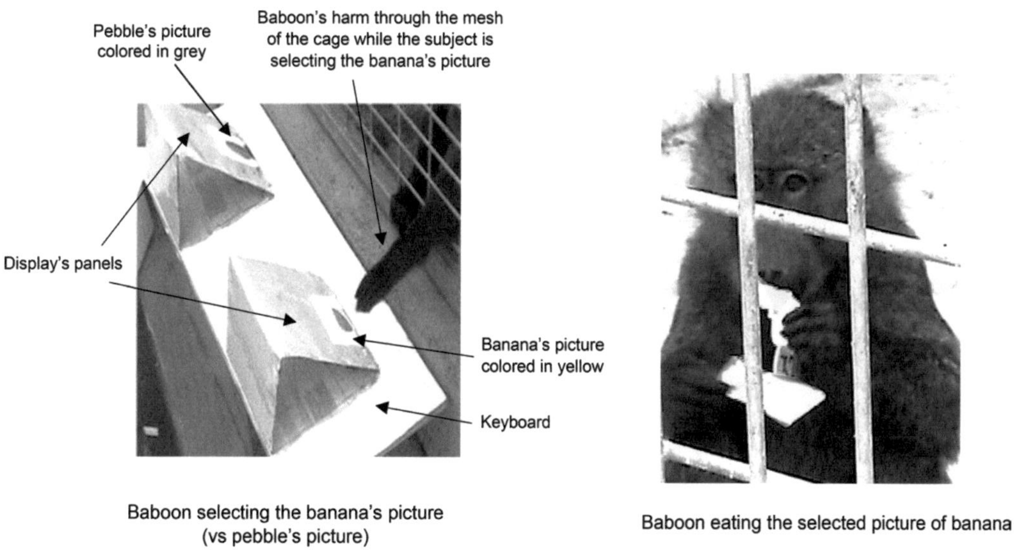

Figure 3. Illustration of the behavioral sequence during test trial 3.

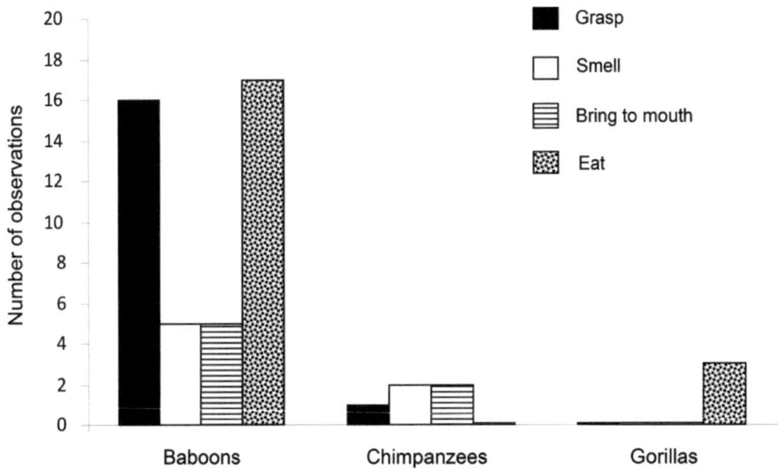

Figure 4. Frequencies of each action observed in baboons, chimpanzees, and gorillas during the first bout of banana picture manipulation, in test trial 3 where subjects had to choose between a banana picture and a pebble picture (illustration from [59]). These results show that baboons massively selected the picture of banana in that test (43 out of the 46 participants). Among them, 16 baboons selected and grasped the picture of banana, 5 smelt it, 5 brought it to mouth, and 17 ate the picture of banana. Similarly, the four gorillas who selected the banana picture ate it. Finally, the five chimpanzees who selected the picture of banana grasped, smelt or brought it to mouth but never ate it.

The demonstration that naive baboons confused real food objects and their picture is in agreement with previous observations from picture processing' experiments in monkeys, for example [57, 58]. The fact that baboons, a species known to fail in self-recognition task [85] and that gorillas, a species less efficient than chimpanzees in MSR, for example [23], both confuse the real object with its picture is consistent with the hypothesis that pictorial ability may correlate MSR skill. In addition, chimpanzees, subjects who frequently succeed in the MSR task, for example [13], showed no confusion in picture processing. Our data might support the hypothesis that the cognitive foundation supporting the use of pictures as referents is more developed in chimpanzees than it is in gorillas and monkeys. Unfortunately, because that research provided no information on MSR of these subjects, it remains unclear whether a relationship actually exists between MSR abilities and iconic symbol use. The second study (unpublished), conducted in collaboration with Joël Fagot and William Hopkins, was thus aimed at alleviating that limitation.

6.2. Picture processing and self-recognition in chimpanzees

Under our hypothesis that self-recognition ability might be based upon the cognitive ability to symbolize, we were expecting that the positive MSR animals would both be able to process their image in the mirror as a representation, and to map that referential stimulus to their internal visual representation of their self. By contrast, the negatively tested animals would be unable to succeed in that mapping process, because the mirror image is processed as a real object, rather than as a representation.

The participants were 32 chimpanzees (*Pan troglodytes*): 9 males and 23 females, from 8 to 22-year-old, all housed in social group of 2–3 individuals in indoor/outdoor enclosures at the Yerkes Regional Primate Research Center. The Yerkes Center is fully accredited by the American

Association for Accreditation of Laboratory Animal Care. American Psychological Association guidelines for the ethical treatment of animals were adhered to during all aspects of this study.

The chimpanzees were proposed two tasks. The first one is a test of MSR inspired from Gallup [2]. The second test is a replica of Parron and collaborators [59] procedure. Conjoint analyses of the findings obtained in these two tests were expected to highlight possible relations between MSR and picture understanding.

6.2.1. The self-recognition task

A camera (with a 2.7 inches LCD screen) taped subjects' behavior when they were facing or not the screen. Each subject received two videotaped sessions, each lasted 3 minutes. The first session was a control condition with the screen turned away from the subject (mirror away condition). That session served for baseline measurement for general self-directed actions. Test session 2 replicated the procedure of session 1, but now with the screen turned toward the subject displaying an image observed by the subject, and therefore used as a "mirror" (mirror condition) (see **Figure 5**).

The coding considered five behavioral categories: "Face-directed actions" defined as manual actions oriented toward the face, such as touching the cheek or the inside of the mouth; "Facial expressions" such as opening the mouth to exhibit the teeth; "Contingent body motions", such as alternate right/left or forward/backward body movements; "others": any behavior different from the previous ones.

When the screen was toward the subjects, a total of 384 actions were recorded for the group, they were distributed as follows: face directed (22.9%); facial expressions (28.6%); contingent body motions (25.5%) and others (45.8%) (see **Figure 6**). Results showed that chimpanzees displayed more actions when the screen was toward them than when it was turned away

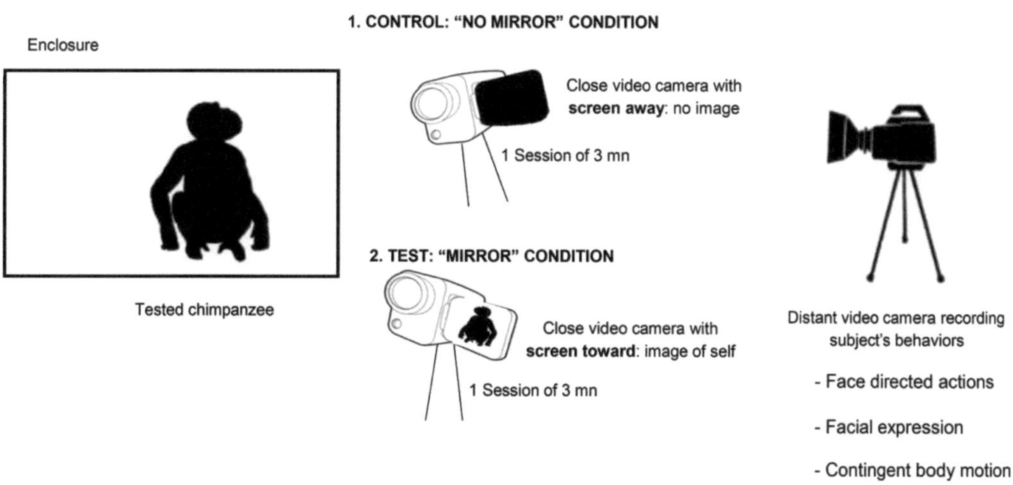

Figure 5. Schematic illustration of the experimental protocol of the self-recognition task in chimpanzees.

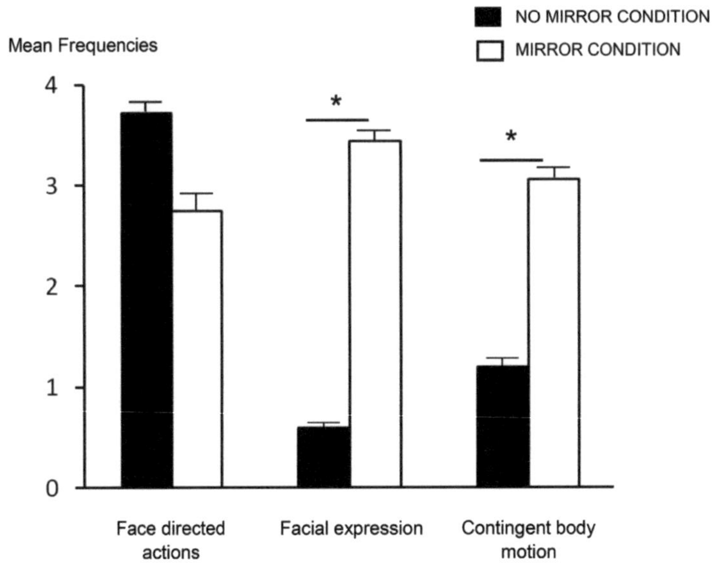

Figure 6. Mean frequencies of each behavioral category observed in chimpanzees depending on the orientation of the screen: away from the subject (no mirror condition), toward the subject (mirror condition). (*Tukey tests, $p < 0.05$). These results globally show that some of the tested chimpanzees displayed some specific behaviors only in front of the mirror, suggesting that these subjects expressed self-recognition. Indeed, even if they did not exhibit more face-directed actions in front of the screen, they nevertheless exhibited more facial expression and contingent body motion in the condition where they could observe themselves: the mirror condition.

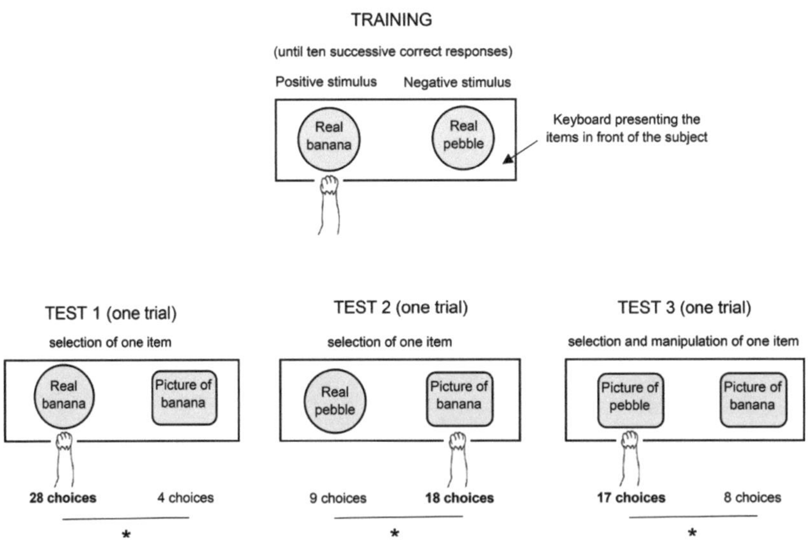

Figure 7. Illustration of the experimental protocol and the frequencies of banana picture choices compared to the other item for the three tests trials in chimpanzees. (*Two-tailed binomial tests, $p < 0.05$). These results show a massive preference for the real banana compared to the picture of banana in test trial 1. When the subjects had to choose between the picture of banana and the real pebble in test trial 2, they clearly exhibited a preference for the picture of banana while they preferred the picture of pebble in test trial 3. Altogether, these mixed results preclude from any conclusion about the picture's preference in this group of chimpanzees. We can notice that all the 32 chimpanzees participated to the first test trial while their number decreased for the two following test trials. It suggests that some subjects may have lost interest in the task when no real food was proposed.

from them ($F(1,31) = 4.90$; $p < 0.05$) and realized an increasing number of facial expressions and contingent body motions (Tukey Honestly Significant Differences test, $p < 0.05$), but not face-directed actions, when the screen was toward them. Although there were no reliable effects for face-directed actions at the group level, four chimpanzees demonstrated an increased frequency of face-directed actions (one-tailed binomial tests, $p < 0.05$) when they saw the screen and thus will be consider as self-recognizer hereafter.

6.2.2. *The picture-understanding task*

The 32 chimpanzees were submitted to the picture task following the same procedure as [59], previously described above.

In test 1, the group had a significant preference for the real banana compared to the banana picture; (two-tailed binomial test, $p < 0.05$). In test 2, 5 chimpanzees expressed no choice, but the rest of the group had a significant preference for the banana picture over the real pebble; (two-tailed binomial test, $p < 0.05$). There was by contrast no preference for the picture of banana at the group level in test 3, and 7 chimpanzees did not choose any item (two-tailed binomial test $p > 0.05$) (see **Figure 7**).

The behavior of the four self-recognizers chimpanzees as determined above was contrasted with that of the 28 non self-recognizers, and showed no reliable difference between the two groups of subjects (Chi-square tests all $ps < 0.05$). As we found no relation between self-recognition

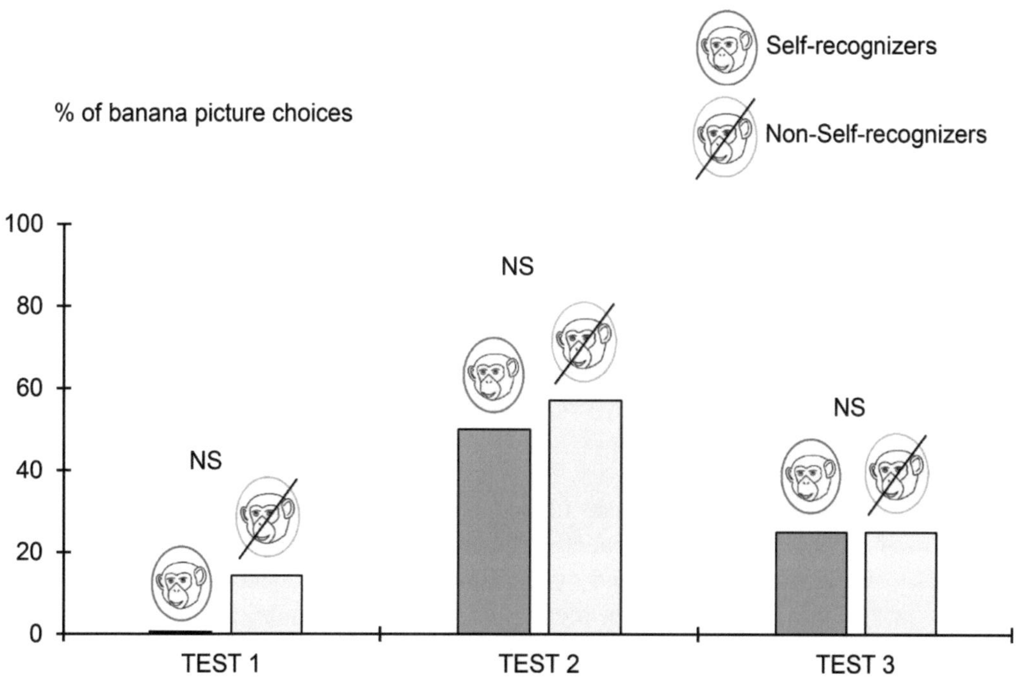

Figure 8. Percentages of banana picture choices in test trials 1–3 of the picture task, according to the self-recognition abilities of the chimpanzees. These results show no difference between self-recognizer and non-self-recognizer subjects as they both selected the banana picture by the same proportion in the three test conditions, independently from their self-recognition abilities. (Chi-square tests all $ps < 0.05$).

	Take	Smell	Bring to mouth	Eat
Self-recognizer	1/4	1/4	0	1/4
Non-self-recognizer	5/28	2/28	4/28	3/28

Table 1. Frequencies of actions observed for the self-recognizer and non-self- recognizer chimpanzees who selected the banana picture in test 3.

and picture processing, it suggests that self-recognizer and non-self-recognizer chimpanzees behaved indistinctively in the picture task, and that the proportion of banana picture selected did not differ as a function of MSR ability (see **Figure 8**).

We also examined the actions performed on the banana picture, once reached in test 3 (see **Table 1**), due to our expectation that only the non-self-recognizer chimpanzees would process the picture as if it was a real object. The number of "take", "smell" and "bring to mouth" actions were minimal, precluding firm conclusion at the group level. Altogether, four instances of eating behaviors were observed, and one of them was expressed by a self-recognizer chimpanzee. That sole observation invalidates our hypothesis that only non-self-recognizer chimpanzees would confuse the actual object and its picture.

7. Discussion

The main originality of our research was to test the hypothesis of a correlation between self-recognition and picture understanding in chimpanzees. Unfortunately, our results did not provide any evidence in support of our assumption that self-recognizer chimpanzees would behave differently with pictures than non-self-recognizer ones. However, these findings should not be taken into account to reject our assumption, as some procedural issues were probably the source of our negative results. On the first hand, our procedure might have been inappropriate to reveal MSR ability, as one chimpanzee who failed to demonstrate self-directed behaviors in our task previously passed the mark test in [13]. It is very likely that the small surface of our camera screen (2.7 inches) might have limited the exploration of the subjects as it did not display the entire subject's reflection. Noticeably, the other studies, assessing self-recognition in primates, used a larger screen (around 15 inches) to display video images, for example [31, 86, 87]. Consequently, the proportion of self-recognizer chimpanzees, particularly small in our study (4/32), might not reflect the reality, regarding the ratios usually reported in the literature, for example, 8/12 in [13]. On the other hand, our picture task might have been irrelevant to test pictorial competence of non-naïve chimpanzees. Overall, chimpanzees, independently of their MSR ability, showed a minimal interest in pictures, which might be explained by the familiarity of these chimpanzees with iconic stimuli, for example [88]. As a result, the low frequencies of manipulations precluded any interpretation of the chimpanzees' behaviors with the banana picture.

Overall, besides our two studies, empirical findings on MSR and pictorial competence in primates are so disparate across species and procedures, that inferring a correlation between both competences from that sole approach appears insufficient. Nevertheless, paralleling the abundant developmental studies in human infants and comparative findings in primates will help in scaffolding a new postulate.

7.1. Developmental course of picture understanding

In their first 2 years of life, children probably miss the symbolic nature of pictures. It is only very gradually that pictorial competence develops, for example [89]. Nine-month-old infants manually explore the pictures as if they were trying to pick-up the depicted objects [83]. They do so even more for realistic pictures compared to non-realistic ones [90]. It is not until 18–24 months of age that children prefer upright to inverted pictures [91] and point at depicted objects rather than manually explore them [83]. In addition, from 24 to 30 months, children can follow a request to put a toy at a place specified to them on a picture and can use information provided with a picture to find the object in the depicted room [92]. Nevertheless, even at age 4, children can show confusion about the properties of pictures and depicted objects [93], and the consequences of actions on pictures and objects [94]. Interestingly, the development of pictorial competence in children fits with the dynamic hypothesis of picture processing in primates described in Section 5. Primates, like infants, begin to process the picture as if it was the real object (confusion mode) and evolves on the high "referential" road of picture processing. They may then achieve the ultimate level, where pictures are processed as referential stimuli (the equivalence mode). Altogether, studies in children and primates suggest that being able to interpret and understand pictures as symbols is a very complex and protracted process.

7.2. Developmental course of mirror self-recognition

7.2.1. The two levels preceding mirror self-recognition

Rochat [95] proposed to divide infants' self-awareness ability into six different levels (from 0 to 6). As disentangling the cognitive abilities of levels 1 and 2 described in [95] is problematic in primates, these two levels will be grouped in only one: the level 1 of "contingency learning". The first two levels of MSR development emerge at the same age in humans and chimpanzees [96]. At level 0, "LEVEL OF CONFUSION", the mirror image is confounded with the reality of the environment it reflects. Indeed, by 4–6 months of age, children treat the mirror image as if it is another child, and primates (monkeys and great apes) display social behavior in front of the mirror. At level 1, "LEVEL OF CONTINGENCY LEARNING", children by 12 months of age begins to search for the image behind the mirror and test contingencies of their movements. Chimpanzees also explore the physical properties of the mirror and begin to detect the contingency between their own bodies and the mirror reflection. Macaques, under certain experimental procedure, are also capable to take their own body as a referential to establish a correspondence between kinesthetic information and external visual effects. They use the mirror to direct their manual searches for otherwise invisible targets [97, 98], to locate and grab an object attached behind their heads [99], or use televised images of their hands to learn to pick-up food [100]. This kinesthetic-visual matching ability warrants the individuals to grasp the correspondence between what the visual image of the body in the mirror looks like and what the body movements feel like [101] and may actually play the role of precursor for self-directed behaviors.

7.2.2. The level of mirror self-recognition

The "LEVEL OF IDENTIFICATION" signifies the emergence of self-directed behaviors. Children by around 15 months of age start passing the mark test; and almost all succeed by 24 months of age [26, 102]. In spite of massive inter-individual differences in MSR ability, it is well documented that chimpanzees develop this ability later in life than humans do, around 6–8 years [46]. Although monkeys systematically fail to pass the mark test, it may be unrelated to their self-recognition ability. It is very likely that, while being able to match kinesthetic and visual information under certain conditions, both species specificity (sensory-motor specializations) and inherent motivation of individuals for interacting with the environment influence monkeys' behaviors displayed in front of a mirror [48]. Indeed, the role of the motivation may be underestimated when comparing the responses of infants, apes and monkeys. Intensive maternal interventions maximize the infants' motivation to explore the mirror and also help to focus their interest in their own face, for example [102]. Somehow, infants are trained to appreciate the connection between themselves and their mirror image [103]. Apes are also naturally motivated to look at the mirror reflection since they are usually more engaged in mutual gaze during social interactions than other primates' species. In contrast, monkeys primarily use the tactile mode to display reciprocal engagement and are more socially inclined to avert direct gaze [104].

From an evolutionary point of view, is there any meaning for individuals of monkeys' society to develop an interest in their *own* facial features? Chang and collaborators [39, 43] designed two studies using some pioneering training procedures to promote the monkeys' motivation to explore their own face. In the first study [39], a visual-somatosensory training helped macaques to interpret the mirror image as their own reflection. While macaques were facing a mirror, an irritating red laser spot was projected onto their face, producing a somatosensory feeling, which encourages them to touch the spot. In order to reinforce the learning process, they also received food reward after successful trials. Following 12–38 days of training, a non-irritating laser was used to project a spot on the macaques' face and food reward was no longer delivered during the test session and subjects yet passed the mark test. They even further spontaneously used the mirror to inspect some hidden parts of their bodies. In their second study [43], Chang and collaborators improved their procedure to exclude criticisms arguing that the extensive training received by monkeys may have promotes some behaviors that merely look-like self-recognition. They used a visual-proprioceptive training to motivate monkeys to locate a spot, visible either directly or through the mirror reflection, projected onto a surface in their close personal space. Macaques further failed to pass the mark test. However, in a second phase, the spot was only visible via the mirror reflection, monkeys, after training, eventually passed the mark test. These pioneering results not only demonstrate that under certain training conditions promoting the contingency learning, monkeys pass the mark test but they also highlight the crucial role of motivation as, unless being compelled to do it, monkeys did not spontaneously use the mirror reflection to touch the projected spot.

7.2.3. Toward a self-concept: the "level of permanence" and the "level of meta-awareness"

Around 3 years of age, infants recognize picture (even taken in the past) and delayed videos of themselves [105, 106], and reach the "LEVEL OF PERMANENCE" [95]. This level signifies that the self is identified beyond the here and now of mirror experience. A permanent self is expressed as invariant over time and appearance changes.

As well as human infants, primates have also been tested with pictures to evaluate their self-recognition ability. Monkeys do not recognize pictorial representations of themselves, for example [27]. By contrast, some studies on chimpanzees or gorillas provide some evidence of self-recognition from pictures. Viki, a home-raised chimpanzee, recognized pictures depicted in books and other materials and imitate actions illustrated in films, photographs, and line drawings. When tested in a categorization task, she was able to sort photographs of chimpanzees and humans into two piles, and not that surprisingly, places her own photography on the human pile [107]. Koko, a sign-language trained gorilla, recognized herself on photograph and even labeled her name on it [108]. Ai, the language-trained chimpanzee, used symbolic labels for individuals, transferred the symbols to label pictures of the individuals, including herself [109]. Remarkably, the sole individuals reported to recognize themselves on pictures, were all intensively trained to associate pictures and objects (from an abundant exposure or a language training) and to process them as some representations. That pictorial competence has probably promoted their ability to recognize themselves on pictures.

Self-recognition can be subdivided into two successive levels: First, the "LEVEL OF IDENTIFICATION", already described above, implies the emergence of self-directed behaviors meaning that individuals manifest recognition and identify that what is in the mirror is "Me". Second, the "LEVEL OF PERMANENCE" refers to a "self-concept", where the self is identified beyond the temporal and spatial animal's experience with a mirror. In that level, self-recognition can occur in absence of a mirror, from delayed videotapes and photographs. This process requires higher cognitive ability that emerges later in the development. This statement is consistent with findings on object permanence in primates. Infant chimpanzees [110] and gorillas [111] are reported to attain the stage 6 in object permanence: that is, these animals are able to take into account the invisible displacements of objects to find objects hidden in successive locations [112]. Monkey species seem to be limited to stage 5 in object permanence [112]: that is, they can only find hidden objects when the displacements are visible [110, 111, 113]. "Level of permanence" and "stage 6 in object permanence" imply a mental representation capacity where the object of interest, the self, is conceived as something permanent in time and space. By contrast, the "level of identification" and "stage 5 in object permanence" might imply a cognitive capacity less demanding, as the object of interest and its representation are simultaneously present in time and space.

Based on the classification of Rochat [95], the ultimate level, to achieve "self-conciousness", is named the "LEVEL OF META SELF-AWARENESS". This level implies that the self is recognized not only from a first person perspective, like in the previous level, but also from a third person's perspective. In other words, individuals are not only aware of what they are (self-concept) but also of how they are in the mind of others (self-consciousness).

Developmental literature in humans and primates shows that cognitive abilities arising from theory of mind actually emerge independently of MSR. In normal children, who pass the mark test between 15 and 24 months [26], complex cognitive abilities such as intentional deception, perspective-taking and empathy develops considerably later in life, for example [114]. In spite of confusing results, it is now acknowledged that apes can show MSR even without being able to attribute intent and emotions in others, as these capacities seem to emerge, if they do, in later phases of their development [115].

Altogether, these results suggest that, through the developmental course of complex cognitive abilities, self-concept may extend far beyond the MSR. Self-recognition demonstrated by self-directed behavior in front of a mirror corresponds to the level of identification while self-awareness, not only requires to achieve first the "level of permanence", but also the subsequent level of "self-conciousness". Moreover, it appears that pictorial competence as implied by an equivalence mode of picture processing may progressively emerge after the development of self-directed behaviors in front of a mirror. Since, MSR may not require self-awareness, then monkeys, who reach stage 5 in object permanence, should be able to achieve the "level of differentiation", that of MSR.

7.3. Pictorial competence and self-recognition correlation: a refined postulate

Empirical findings and theoretical proposals of the literature have contributed to support and refine my initial assumption that a dynamic pictorial competence correlates self-recognition ability. This model of correlation (see **Figure 9**) is based on the ontogenetic sequence of cognitive abilities observed in infants, and comparative findings in primates. I acknowledge that establishing a such correlation across species is somehow problematic as it would require comparability in procedures and in assessment. Nevertheless, I believe that, even if the time scale and the quantitative nature of cognitive processes may differ from humans to great apes and from great apes to monkeys, this model may open up new perspectives as it is the first attempt to correlate symbolic cognitive ability to embodied cognition and to self-recognition competence.

The starting point of this model is the level of "CONFUSION": the animals mistake the two-dimensional stimulus (pictures or mirror image) with the real object represented (an object or their self). Then, following a growing experience in "ASSOCIATIVE LEARNING", animals learn both visual-visual matching and kinesthetic-visual matching. The former ability allows the animals to establish an association between the picture and the real object depicted, and the latter to use the mirror to locate objects that are outside their direct visual field. After repeated exposure to the stimulus (picture or mirror), if animals habituate to the stimulus, they may follow the low "perceptual" road of cognitive processing which corresponds to the level of "INDEPENDENCE". Pictures are then processed as a combination of physical features, regardless of their referential content, and the mirror is perceived as a common object, with no reflective properties. More likely, the extensive phase of associative learning allows animals to reach the level of "IDENTIFICATION" which corresponds to the cognitive ability to identify object from representation. Although animals progress on the high "referential" road of cognitive processing, they do not yet fully grasp the referential nature of picture. This level also corresponds to MSR ability, which certainly required contingency learning but not a complex representational capacity related to self-awareness. With the development of more complex cognitive abilities, which remain to be determined, animals reach the "LEVEL OF REPRESENTATION". By forming a mental representation of the real object, animals are now capable to compare this mentally held information to the external visual stimulus. Therefore, they are able to interpret the picture as a representation of a real object (equivalence mode

http://dx.doi.org/10.5772/intechopen.75568

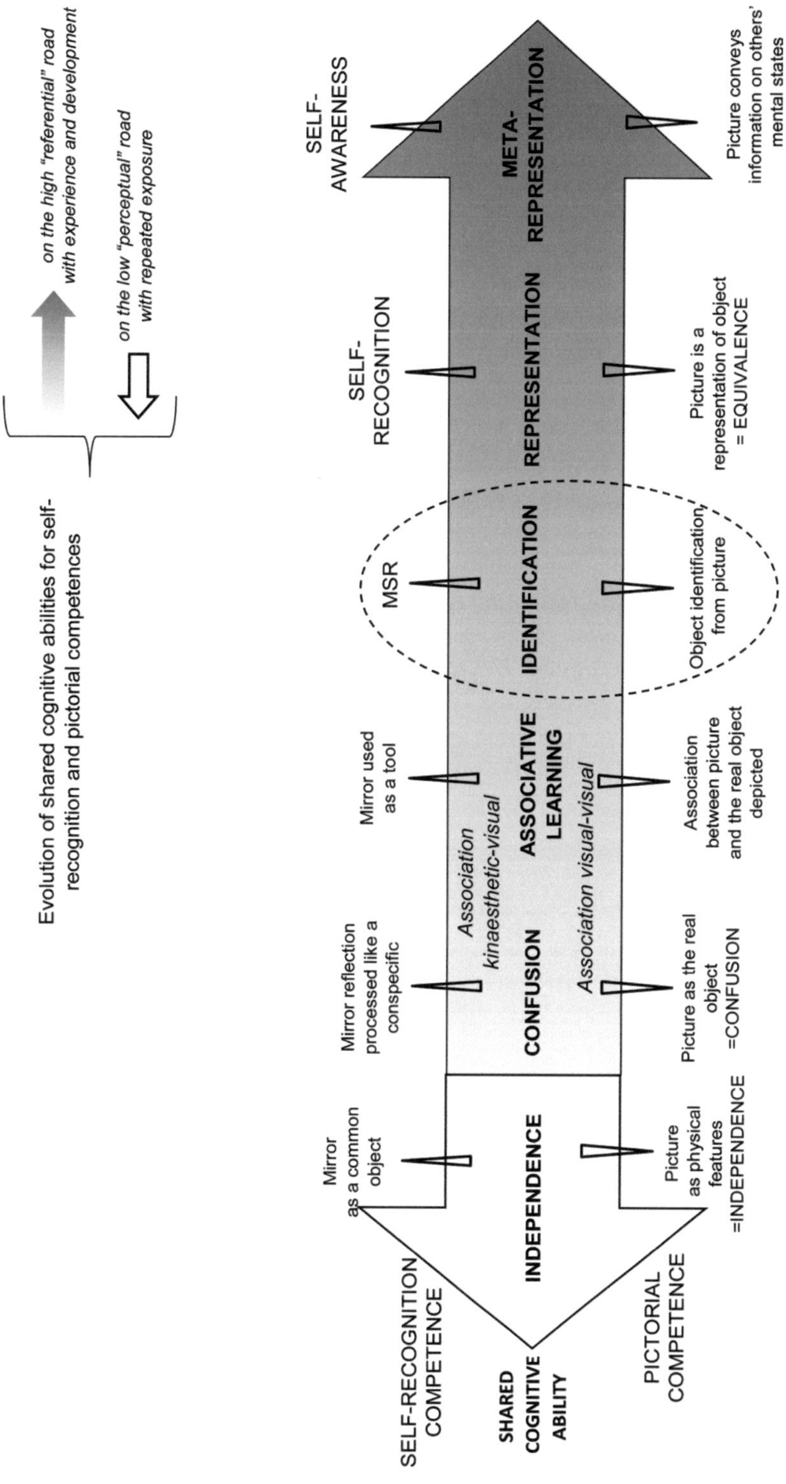

Figure 9. A conceptual model of the developmental course of shared cognitive abilities underlying self-recognition and picture processing in primates.

of picture processing), and to recognize themselves from more ambiguous medium such as pictures. Finally, the ultimate "LEVEL OF META-REPRESENTATION" corresponds to the development of cognitive abilities, which seems to follow from theory of mind. At this stage, a self-awareness concept is expected to emerge. Regarding the pictorial competence, one can hypothesize that animals might be able to interpret symbolic medium to understand other's minds and to transfer this knowledge to the real world.

7.4. Conclusion

My original hypothesis, that being able of self-recognition might be related to the capacity of processing picture as a symbolic representation of the real object, is consistent with this model of cognitive correlation. The following main outcomes may be drawn up. First, the pictorial competence correlating MSR cognitive stage may be less complex than that corresponding to the equivalence mode of processing. MSR seems rather co-develops with an extensive experience in associating picture and object until the grasping of their correspondence. As monkeys are reported to possess this associative cognitive ability, they should be able to display MSR. It therefore suggests that monkeys' repetitive failures to pass the mark test may be unrelated to self-recognition ability *per se,* and more likely due to the procedural constraints of this test. Second, due to over interpretations of MSR responses, the strong statement that skills related to self-recognition have evolved until a cognitive gap in the phylogeny has been spread. This assumption may be partially left unsupported as, according to this model of cognitive correlation (**Figure 9**), both great apes and monkeys achieve the same level of cognitive capacities at the MSR stage. This observation should encourage investigations on the origin of such MSR responses discrepancies in the literature. Third, the traditional interpretation of MSR as an index of self-awareness [4, 5] raises serious doubts based on the ontogenetic emergence of MSR. Indeed, the development of complex cognitive abilities related to self-awareness occurs much later than MSR ability. Finally, MSR is a quite good illustration of how inappropriate experimental procedure, may lead to erroneous interpretations on the animals' cognitive abilities. The fact mirror self-recognition is so automatic and effortless for most of us that has probably encouraged an anthropocentric bias in the design of experimental procedure. Swartz in [116] already underlined how "a single method of measurement may contain sources of measurement error that are unique to that method". Yet, multiplying the methods of measurement would be unfruitful if the construction of the tests is still anthropocentrically based and keeps on favoring the human-like species. To conclude, future reflection on the experimental approach in comparative cognition, which is still heavily anthropocentric, may help to move away from our preconceptions and would certainly shed new light on animal cognitive mechanisms that were, until then, confined to humans. Would this perspective help researchers working on comparative cognition to live happily ever after? To be continued….

Acknowledgements

I would like to thank the animal care and the facilities teams of the Yerkes National Primate Research Center in Atlanta, the Wolfgang Kohler Primate Center at the Leipzig Zoo, the

http://dx.doi.org/10.5772/intechopen.75568

Nuremberg Zoo, and the station of primatology of Rousset, for their support. I would like to thank Joel Fagot, Josep Call and William Hopkins for their hospitality and complete cooperation in the experiments. I am very grateful to Marie Montant for providing her smart comments, and to Ethan Conia, of the Saint-Claude facility team, for his help in the finalization of the chapter.

Author details

Parron Carole

Address all correspondence to: carole.parron@univ-amu.fr

Laboratory of Cognitive Psychology, National Center of Scientific Research (CNRS)-Université d'Aix-Marseille, France

References

[1] Morin A, DeBlois S. Gallup's mirrors: More than an operationalization of self-awareness in primates? Psychological Reports. 1989;**65**(1):287-291

[2] Gallup GG. Chimpanzees: Self-recognition. Science. 1970;**167**(3914):86-87

[3] Gallup GG. Mirror-image stimulation. Psychological Bulletin. 1968;**70**:782-793

[4] Gallup GG. Self-awareness and the emergence of mind in primates. American Journal of Primatology. 1982;**2**(3):237-248

[5] Gallup GG. Self-awareness and the evolution of social intelligence. Behavioural Processes. 1998;**42**(2):239-247

[6] Premack D, Woodruff G. Does the chimpanzee have a theory of mind? Behavioral and Brain Sciences. 1978;**1**(4):515-526

[7] Savage-Rumbaugh ES, Rumbaugh DM, Boysen S. Symbolic communication between two chimpanzees (*Pan troglodytes*). Science. 1978;**201**(4356):641-644

[8] Povinelli DJ, Eddy TJ. Chimpanzees: Joint visual attention. Psychological Science. 1996;**7**(3):129-135

[9] Swartz KB. Self-recognition in nonhuman primates. In: Greenberg G, Haraway M, editors. Comparative Psychology: A Handbook. New York: Garland; 1998. pp. 849-855

[10] Hart D, Karmel MP. Self-awareness and self-knowledge in humans, apes, and monkeys. In: Russon AE, Bard KA, Parker ST, editors. Reaching into Thought: The Minds of the Great Apes. New York: Cambridge University Press; 1996. pp. 325-347

[11] Bard KA, Todd BK, Bernier C, Love J, Leavens DA. Self-awareness in human and chimpanzee infants: What is measured and what is meant by the mark and mirror test? Infancy. 2006;**9**(2):191-219

[12] Anderson JR, Gallup GG. Self-recognition in nonhuman primates: Past and future challenges. In: Haug M, Whalen RF, editors. Animal Models of Human Emotion and Cognition. Washington, DC: American Psychological Association; 1999. pp. 175-194

[13] Lin AC, Bard KA, Anderson JR. Development of self-recognition in chimpanzees (*Pan troglodytes*). Journal of Comparative Psychology. 1992;**106**(2):120-127

[14] Hyatt CW, Hopkins W. Self-awareness in bonobos and chimpanzees: A comparative perspective. In: Parker ST, Mitchell R, Boccia ML, editors. Self-Awareness in Animals and Humans: Developmental Perspectives. Cambridge University Press New-York. 1994. pp. 248-253

[15] Suárez SD, Gallup GG. Self-recognition in chimpanzees and orangutans, but not gorillas. Journal of human evolution. 1981;**10**(2):175-188

[16] Swartz KB, Evans S. Social and cognitive factors in chimpanzee and gorilla mirror behavior and self-recognition. In: Parker ST, Mitchell R, Boccia ML, editors. Self-Awareness in Animals and Humans: Developmental Perspectives. New-York: Cambridge University Press;1994. pp. 189-206

[17] Ledbetter DH, Basen JA. Failure to demonstrate self-recognition in gorillas. American Journal of Primatology. 1982;**2**(3):307-310

[18] Parker ST. Incipient mirror self-recognition in zoo gorillas and chimpanzees. In: Parker ST, Mitchell R, Boccia ML, editors. Self-Awareness in Animals and Humans: Developmental Perspectives. New-York: Cambridge University Press; 1994. pp. 301-307

[19] Patterson FGP, Cohn RH. Self-recognition and self-awareness in Lowland Gorillas. In: Parker ST, Mitchell R, Boccia ML, editors. Self-Awareness in Animals and Humans: Developmental Perspectives. New-York: Cambridge University Press; 1994. pp. 273-290

[20] Posada S, Colell M. Another gorilla (*Gorilla gorilla gorilla*) recognizes himself in a mirror. American Journal of Primatology. 2007;**69**(5):576-583

[21] Allen M, Schwartz BL. Mirror self-recognition in a gorilla (*Gorilla gorilla gorilla*). Journal of Integrative Biosciences. 2008;**5**:19-24

[22] Nicholson IS, Gould JE. Mirror mediated object discrimination and self-directed behavior in a female gorilla. Primates. 1995;**36**(4):515-521

[23] Shillito DJ, Gallup GG, Beck BB. Factors affecting mirror behaviour in western lowland gorillas, Gorilla gorilla. Animal Behaviour. 1999;**57**(5):999-1004

[24] Shumaker RW, Swartz KB. When traditional methodologies fail: Cognitive studies of great apes. In: Bekoff M, Allen C, Burghardt GM, editors. The Cognitive Animal: Empirical and Theoretical Perspectives on Animal Cognition. Cambridge, MA: MIT Press; 2002. pp. 335-343

[25] Anderson JR, Gallup GG. Which primates recognize themselves in mirrors? PLoS Biology. 2011;**9**(3):e1001024

[26] Anderson JR. The development of self-recognition: A review. Developmental Psychobiology. 1984;**17**(1):35-49

[27] Anderson JR. The monkey in the mirror: A strange conspecific. In: Parker ST, Mitchell R, Boccia ML, editors. Self-Awareness in Animals and Humans: Developmental Perspectives. New-York: Cambridge University Press; 1994. pp. 315-329

[28] Suarez SD, Gallup GG. Social responding to mirrors in rhesus macaques (*Macaca mulatta*): Effects of changing mirror location. American Journal of Primatology. 1986;**11**(3):239-244

[29] Gallup GG, Suarez SD. Social responding to mirrors in rhesus monkeys (*Macaca mulatta*): Effects of temporary mirror removal. Journal of Comparative Psychology. 1991;**105**(4):376-379

[30] Neiworth JJ, Anders SL, Parsons RR. Tracking responses related to self-recognition: A frequency comparison of responses to mirrors, photographs, and videotapes by cotton top tamanins (*Saguinus oedipus*). Journal of Comparative Psychology. 2001;**15**(4):432-438

[31] Anderson JR, Kuroshima H, Paukner A, Fujita K. Capuchin monkeys (*Cebus apella*) respond to video images of themselves. Animal Cognition. 2009;**12**(1):55-62

[32] Heschl A, Burkart J. A new mark test for mirror self-recognition in non-human primates. Primates. 2006;**47**(3):187-198

[33] Roma PG, Silberberg A, Huntsberry ME, Christensen CJ, Ruggiero AM, Suomi SJ. Mark tests for mirror self-recognition in capuchin monkeys (*Cebus apella*) trained to touch marks. American Journal of Primatology. 2007;**69**(9):989-1000

[34] Anderson JR, Roeder JJ. Responses of capuchin monkeys (*Cebus apella*) to different conditions of mirror-image stimulation. Primates. 1989;**30**(4):581-587

[35] Paukner A, Anderson JR, Fujita K. Reactions of capuchin monkeys (*Cebus apella*) to multiple mirrors. Behavioural Processes. 2004;**66**(1):1-6

[36] Macellini S, Ferrari PF, Bonini L, Fogassi L, Paukner A. A modified mark test for own-body recognition in pig-tailed macaques (Macaca nemestrina). Animal Cognition. 2010;**13**(4):631-639

[37] Rajala AZ, Reininger KR, Lancaster KM, Populin LC. Rhesus monkeys (*Macaca mulatta*) do recognize themselves in the mirror: Implications for the evolution of self-recognition. PLoS One. 2010;**5**(9):e12865

[38] Anderson JR, Gallup GG. Do rhesus monkeys recognize themselves in mirrors? American Journal of Primatology. 2011;**73**(7):603-606

[39] Chang LJ, Gianaros PJ, Manuck SB, Krishnan A, Wager TD. A sensitive and specific neural signature for picture-induced negative affect. PLoS Biology. 2015;**13**(6):e1002180

[40] Toda K, Platt ML. Animal cognition: Monkeys pass the mirror test. Current Biology. 2015;**25**(2):R64-R66

[41] Huttunen AW, Adams GK, Platt ML. Can self-awareness be taught? Monkeys pass the mirror test-again. Proceedings of the National Academy of Sciences. 2017;**114**(13):3281-3283

[42] Anderson JR, Gallup GG. Mirror self-recognition: A review and critique of attempts to promote and engineer self-recognition in primates. Primates. 2015;**56**(4):317-326

[43] Chang L, Zhang S, Poo MM, Gong N. Spontaneous expression of mirror self-recognition in monkeys after learning precise visual-proprioceptive association for mirror images. Proceedings of the National Academy of Sciences. 2017;**114**(12):3258-3263

[44] Swartz KB. What is mirror self-recognition in nonhuman primates, and what is it not? Annals of the New York Academy of Sciences. 1997;**818**(1):65-71

[45] Swartz KB, Evans S. Not all chimpanzees (*Pan troglodytes*) show self-recognition. Primates. 1991;**32**(4):483-496

[46] Povinelli DJ, Rulf AB, Landau KR, Bierschwale DT. Self-recognition in chimpanzees (*Pan troglodytes*): Distribution, ontogeny, and patterns of emergence. Journal of Comparative Psychology. 1993;**107**(4):347-372

[47] Gallup GG. Self-recognition: Research strategies and experimental design. In: Parker ST., Mitchell R, Boccia L., editors. Self-Awareness in Animals and Humans: Developmental Perspectives. New-York: Cambridge University Press; 1994. pp. 35-50

[48] De Veer MW, van den Bos R. A critical review of methodology and interpretation of mirror self-recognition research in nonhuman primates. Animal Behaviour. 1999;**58**:459-468

[49] Yamagiwa J. Functional analysis of social staring behavior in an all-male group of mountain gorillas. Primates. 1992;**33**(4):523-544

[50] De Waal FBM, Luttrell LM. Toward a comparative socioecology of the genus *Macaca:* Different dominance styles in rhesus and stumptail monkeys. American Journal of Primatology. 1989;**19**:83-109

[51] De Waal FBM. Chimpanzee Politics: Power and Sex among Apes. New York: Harper & Row; 1982. 223 p

[52] Heyes CM. Social learning in animals: Categories and mechanisms. Biological Reviews. 1994;**69**(2):207-231

[53] Heyes CM. Self-recognition in primates: Further reflections create a hall of mirrors. Animal Behaviour. 1995;**50**(6):1533-1542

[54] Fagot J, Martin-Malivel J, Dépy D. What is the evidence for an equivalence between objects and pictures in birds and nonhuman primates? In: Fagot J, editor. Picture Perception in Animals. Psychology Press Ltd, East Sussex; 2000. pp. 295-320

[55] Perrett DJ, Mistlin AJ. Perception of facial characteristics by monkeys. In: Stebbins WC, Berkley MA, editors. Comparative Perception. Vol. II. Complex Signals. New York: John Wiley & Sons; 1990. pp. 187-213

[56] Wright PC. The nocturnal primate niche in the new world. Journal of Human Evolution. 1989;**18**(7):635-658

[57] Sackett GP. Monkeys reared in isolation with pictures as visual input: Evidence for an innate releasing mechanism. Science. 1966;**154**(3755):1468-1473

[58] Bovet D, Vauclair J. Functional categorization of objects and of their pictures in baboons (*Papio anubis*). Learning and Motivation. 1998;**29**(3):309-322

[59] Parron C, Call J, Fagot J. Behavioural responses to photographs by pictorially naive baboons (*Papio anubis*), gorillas (*Gorilla gorilla*) and chimpanzees (*Pan troglodytes*). Behavioural Processes. 2008;**78**(3):351-357

[60] D'amato MR, Van Sant P. The person concept in monkeys (*Cebus apella*). Journal of Experimental Psychology: Animal Behavior Processes. 1988;**14**(1):43-56

[61] Martin-Malivel J, Fagot J. Perception of pictorial human faces by baboons: Effects of stimulus orientation on discrimination performance. Learning & Behavior. 2001;**29**(1):10-20

[62] Martin-Malivel J, Mangini MC, Fagot J, Biederman I. Do humans and baboons use the same information when categorizing human and baboon faces? Psychological Science. 2006;**17**(7):599-607

[63] Winner E, Ettlinger G. Do chimpanzees recognize photographs as representations of objects? Neuropsychologia. 1979;**17**(3):413-420

[64] Herrmann E, Melis AP, Tomasello M. Apes' use of iconic cues in the object-choice task. Animal Cognition. 2006;**9**(2):118

[65] Pokorny JJ, de Waal FB. Monkeys recognize the faces of group mates in photographs. Proceedings of the National Academy of Sciences. 2009;**106**(51):21539-21543

[66] Pokorny JJ, de Waal F. Face recognition in capuchin monkeys (*Cebus apella*). Journal of Comparative Psychology. 2009;**123**(2):151-160

[67] Micheletta J, Whitehouse J, Parr LA, Waller BM. Facial expression recognition in crested macaques (*Macaca nigra*). Animal Cognition. 2015;**18**(4):985-990

[68] Truppa V, Spinozzi G, Stegagno T, Fagot J. Picture processing in tufted capuchin monkeys (*Cebus apella*). Behavioural Processes. 2009;**82**(2):140-152

[69] Rumbaugh DM, Warner H, Glasersfeld E. The Lana project: Origin and tactics. In: Rumbaugh DM, editor. Language Learning by a Chimpanzee. New York: Academic Press; 1977. pp. 87-90

[70] Savage-Rumbaugh ES, Rumbaugh DM, Smith ST, Lawson J. Reference: The linguistic essential. Science. 1980;**210**(4472):922-925

[71] Savage-Rumbaugh ES. Ape Language: From Conditioned Response to Symbol. New York: Columbia University Press; 1986. 433 p

[72] Matsuzawa T. The Ai project: Historical and ecological contexts. Animal Cognition. 2003;**6**(4):199-211

[73] Matsuzawa T. Form perception and visual acuity in a chimpanzee. Folia Primatologica. 1990;**55**(1):24-32

[74] Itakura S. Recognition of line-drawing representations by a chimpanzee (*Pan troglodytes*). The Journal of general psychology. 1994;**121**(3):189-197

[75] Tanaka M. Development of the visual preference of chimpanzees (*Pan troglodytes*) for photographs of primates: Effect of social experience. Primates. 2007;**48**(4):303-309

[76] Bartus RT, Dean III RL, Fleming DL. Aging in the rhesus monkey: Effects on visual discrimination learning and reversal learning. Journal of Gerontology. 1979;**34**(2):209-219

[77] Mell T, Heekeren HR, Marschner A, Wartenburger I, Villringer A, Reischies FM. Effect of aging on stimulus-reward association learning. Neuropsychologia. 2005;**43**(4):554-563

[78] Close J, Call J. From colour photographs to black-and-white line drawings: An assessment of chimpanzees' (*Pan troglodytes*) transfer behaviour. Animal Cognition. 2015;**18**(2):437-449

[79] Malone DR, Tolan JC, Rogers CM. Cross-modal matching of objects and photographs in the monkey. Neuropsychologia. 1980;**18**(6):693-697

[80] Tolan JC, Rogers CM, Malone DR. Cross-modal matching in monkeys: Altered visual cues and delay. Neuropsychologia. 1981;**19**(2):289-300

[81] Humphrey NK. Vision in a monkey without striate cortex: A case study. Perception. 1974;**3**(3):241-255

[82] Fagot J, Thompson RK, Parron C. How to read a picture: Lessons from nonhuman primates. Proceedings of the National Academy of Sciences. 2010;**107**(2):519-520

[83] DeLoache JS, Pierroutsakos SL, Uttal DH, Rosengren KS, Gottlieb A. Grasping the nature of pictures. Psychological Science. 1998;**9**(3):205-210

[84] DeLoache JS. Becoming symbol-minded. Trends in Cognitive Sciences. 2004;**8**(2):66-70

[85] Benhar EE, Carlton PL Samuel D. A search for mirror-image reinforcement and self-recognition in the baboon. In: Kondo S, Kawai M, Ehara S, editors. Contemporary Primatology: Proceedings of the 5th International Congress of Primatology; New York: Karger; 1975. pp. 202-298

[86] Menzel EW, Savage-Rumbaugh ES, Lawson J. Chimpanzee (*Pan troglodytes*) spatial problem solving with the use of mirrors and televised equivalents of mirrors. Journal of Comparative Psychology. 1985;**99**(2):211-217

[87] Law LE, Lock AJ. Do gorillas recognize themselves on television. In: Parker ST, Mitchell R, Boccia ML, editors. Self-Awareness in Animals and Humans: Developmental Perspectives. New-York: Cambridge University Press; 1994. pp. 308-312

[88] Parr LA, Winslow JT, Hopkins WD, de Waal F. Recognizing facial cues: Individual discrimination by chimpanzees (*Pan troglodytes*) and rhesus monkeys (*Macaca mulatta*). Journal of Comparative Psychology. 2000;**114**(1):47-60

[89] Troseth GL, Pierroutsakos SL, DeLoache JS. From the innocent to the intelligent eye: The early development of pictorial competence. Advances in Child Development and Behavior. 2004;**32**:1-35

[90] Pierroutsakos SL, DeLoache JS. Infants' manual exploration of pictorial objects varying in realism. Infancy. 2003;**4**(1):141-156

[91] DeLoache JS, Uttal DH, Pierroutsakos SL. What's up? The development of an orientation preference for picture books. Journal of Cognition and Development. 2000;**1**(1):81-95

[92] DeLoache JS, Burns NM. Early understanding of the representational function of pictures. Cognition. 1994;**52**(2):83-110

[93] Robinson EJ, Nye R, Thomas GV. Children's conceptions of the relationship between pictures and their referents. Cognitive Development. 1994;**9**(2):165-191

[94] Flavell JH, Flavell ER, Green FL, Korfmacher JE. Do young children think of television images as pictures or real objects? Journal of Broadcasting & Electronic Media. 1990;**34**(4):399-419

[95] Rochat P. Five levels of self-awareness as they unfold early in life. Consciousness and Cognition. 2003;**12**(4):717-731

[96] Robert S. Ontogeny of mirror behavior in two species of great apes. American Journal of Primatology. 1986;**10**(2):109-117

[97] Anderson JR. Mirror-mediated finding of hidden food by monkeys (*Macaca tonkeana* and *M. fascicularis*). Journal of Comparative Psychology. 1986;**100**(3):237-242

[98] Itakura S. Mirror guided behavior in Japanese monkeys (*Macaca fuscata fuscata*). Primates. 1987;**28**(2):149-161

[99] Itakura S. Use of a mirror to direct their responses in Japanese monkeys (*Macaca fuscata fuscata*). Primates. 1987;**28**(3):343-352

[100] Iriki A, Tanaka M, Obayashi S, Iwamura Y. vSelf-images in the video monitor coded by monkey intraparietal neurons. Neuroscience Research. 2001;**40**(2):163-173

[101] Mitchell RW. Kinesthetic-visual matching, imitation, and self-recognition. In: Bekoff M, Allen C, Burghardt G, editors. The Cognitive Animal. Cambridge, MA: MIT Press; 2002. pp. 345-351

[102] Amsterdam B. Mirror self-image reactions before age two. Developmental Psychobiology. 1972;**5**(4):297-305

[103] Nielsen M, Dissanayake C, Kashima Y. A longitudinal investigation of self-other discrimination and the emergence of mirror self-recognition. Infant Behavior and Development. 2003;**26**(2):213-226

[104] Bard KA, Myowa-Yamakoshi M, Tomonaga M, Tanaka M, Costall A, Matsuzawa T. Group differences in the mutual gaze of chimpanzees (*Pan troglodytes*). Developmental Psychology. 2005;**41**(4):616-624

[105] Povinelli DJ. The unduplicated self. In the self. In: Rochat P, editor. Early Infancy. Amsterdam, The Netherlands: North-Holland-Elsevier; 1995. pp. 162-192

[106] Povinelli DJ. The self: Elevated in consciousness and extended in time. In: Moore C, Lemmon K, editors. The Self in Time: Developmental Perspectives. Mahwah, New Jersey: Lawrence Erlbaum Associates, Inc.; 2001. pp. 79-96

[107] Hayes KJ, Nissen CH. Higher mental functions of a home-raised chimpanzee. Behavior of Nonhuman Primates: Modern Research Trends. 1971;**4**:59-115

[108] Patterson FG. Conversations with a gorilla. National Geographic. 1978;**154**:438-465

[109] Itakura S. A chimpanzee with the ability to learn the use of personal pronouns. The Psychological Record. 1992;**42**(2):157-172

[110] Wood S, Moriarty KM, Gardner BT, Gardner RA. Object permanence in child and chimpanzee. Learning & Behavior. 1980;**8**(1):3-9

[111] Natale F, Antinucci F, Spinozzi G, Potí P. Stage 6 object concept in nonhuman primate cognition: A comparison between gorilla (*Gorilla gorilla gorilla*) and Japanese macaque (*Macaca fuscata*). Journal of Comparative Psychology. 1986;**100**(4):335-339

[112] Piaget J. The Construction of Reality in the Child. New York: Basic Books; 1954 p. 386

[113] Natale F, Antinucci F. Stage 6 object-concept and representation. In: Antinucci F, editor. Cognitive Structure and Development in Nonhuman Primates. Hillsdale (NJ): Lawrence Erlbaum; 1989. pp. 97-112

[114] Gopnik A, Meltzoff AN. Minds, bodies and persons: Young children's understanding of the self and others as reflected in imitation and "theory of mind" research. In: Parker ST, Mitchell R, Boccia ML, editors. Self-Awareness in Animals and Humans: Developmental Perspectives. Cambridge University Press New-York. 1994. pp. 166-186

[115] Povinelli DJ. What chimpanzees (might) know about the mind. In: Wrangham RW, McGrew WC, de Waal FBM, Heltne PG, editors. Chimpanzee cultures, Cambridge, MA: Harvard University Press. 1994. pp. 285-300

[116] Swartz KB. The concept of mind in comparative psychology. Annals of the New York Academy of Sciences. 1990;**602**:105-111

Chapter 7

The Origins of Gibbon Ape Leukaemia Virus

Gregory Stewart Simmons and Gervais Habarugira

Additional information is available at the end of the chapter

http://dx.doi.org/10.5772/intechopen.71694

Abstract

Gibbon ape leukaemia virus (GALV) was first isolated in the early 1970s after a number of gibbons that were housed at the SEATO medical research in Bangkok, Thailand, were diagnosed with lymphoid tumours including malignant lymphoma. It is a novel gamma retrovirus that has never been isolated from wild gibbons. It appears that GALV occurred as a result of a species jump from another as yet unidentified vertebrate host. The full sequence of GALV suggests that it is related loosely to murine leukaemia viruses and a number of rodent species from Southeast Asia have been suggested as possible hosts of the ancestor to GALV. However, no proviral sequence from any Southeast Asian vertebrate has been so far isolated which could be a candidate virus. More recently, two closely related viruses have been found in koalas and a native Australian rat, the grassland melomys (*Melomys burtoni*). These are koala retrovirus (KoRV) and *Melomys burtoni* retrovirus (MbRV). A number of theories have been published recently which endeavour to explain the origins of GALV and its relationship to other viruses including KoRV. Here, the history of GALV is documented and the strengths and weaknesses of current theories on the origin of this virus are discussed.

Keywords: gibbon ape leukaemia virus, koala retrovirus, *Melomys burtoni* retrovirus

1. Introduction

Retroviruses are a unique group of viruses, which have evolved a novel reproductive strategy. They are single-stranded positive-sense, non-segmented RNA viruses which use a unique enzyme, reverse transcriptase to turn their RNA back into DNA, hence the name "retro" virus [1]. Once this is achieved, they use a second enzyme, integrase to insert the viral cDNA into the infected cells chromosomes [2]. This chromosomal DNA is then transcribed and translated to make new viral RNA and proteins, which are then assembled into new virions [3] (**Figure 1**). The viral DNA is referred to as the provirus. Ordinarily, this provirus is inserted into chromosomes of somatic cells, often lymphocytes or their precursors. These viruses are infectious, transmitted horizontally from one individual to another and are said to be exogenous [4]. Should the provirus become inserted into chromosomes in a germ line cell (sperm or ova) there is the potential for the provirus to be passed from parent to offspring and be inherited as with other genes. In this specific scenario, the virus is now said to be endogenous and it is transmitted vertically [5, 6]. Almost all endogenous retroviruses are ancient and considered to be "viral fossils" or the relics of ancient infections [7, 8]. The proviral DNA in these circumstances has often incurred fatal mutations over time and is now no longer capable of producing infectious virions. This proviral DNA is part of the non-coding portion of the host's chromosomes, sometimes referred to as "junk" DNA [9, 10].

Currently, there are seven genera of retroviruses based on their genetic organisation. Alpharetroviruses, betaretroviruses and gammaretroviruses are simple retroviruses for which the genome encodes three genes, *gag*, *pol*/*pro* and *env*. Deltaretroviruses, epsilonretroviruses, lentiviruses and spumaviruses are more complex retroviruses, with a genome

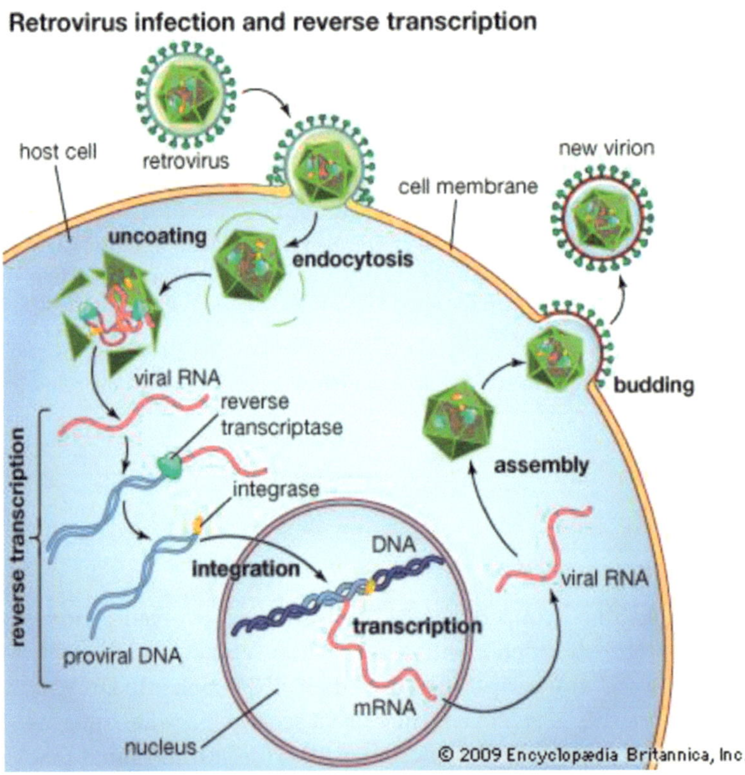

Figure 1. Following retrovirus infection, reverse transcriptase converts viral RNA into proviral DNA, which, by the action of integrase, is then integrated (incorporated) into the DNA of the host cell. This integrated proviral DNA is then transcribed and translated to give new viral RNA and proteins that later give new virions. (From Britannica and King [11], Encyclopaedia Britannica).

encoding additional small regulatory proteins. Gamma retroviruses, formally "type C" retroviruses, have a simple genome approximately 8–10,000 bases long [12].

Retroviruses commonly cause disease by one of two mechanisms. They are often oncogenic and induce cancers by disruption of normal cellular control mechanisms [13, 14]. Many also induce immunodeficiency leading to opportunistic infections in the host [15, 16].

2. History of gibbon ape leukaemia virus

In the mid 1960s, the SEATO medical research facility, now the Armed Forces Institute of Medical Sciences (AFRIMS), was established in Bangkok, Thailand. This organisation conducted medical research on a number of diseases and used a number of different animal species for their research [17, 18]. A number of SEATO annual progress reports, available at (http://www.afrims.org/weblib/apr/aprF.shtml), shed light on some of those early experiments. In 1965, the first white-handed gibbons (*Hylobates lar*) were acquired and the colony grew over the following years [19]. These gibbons were used for research on a range of diseases but a major focus appears to have been malaria and dengue fever virus (DFV). In1969, it was reported that four gibbons over a 20-month period were diagnosed with malignant lymphoma [20, 21]. As this seemed to be an unusually high incidence of an uncommon disease, endeavours were made to look for a viral aetiology. Prior to this episode, there had been very few reports of lymphoid tumours in gibbons or primates other than man [22, 23].

Subsequently, a novel gamma retrovirus was isolated and it was given the name gibbon ape leukaemia virus (GALV) [24]. In the following years, a number of strains of GALV were isolated and the full sequence of their genetic codes has now been published [25]. Gibbon ape leukaemia virus is an infectious exogenous gammaretrovirus. Currently, there are four strains, which have been isolated from gibbons. These are GALV-SEATO for the strain detected at the AFRIMS (SEATO) facility [24], GALV-H from a colony of gibbons kept at Halls Island, Bermuda [26], GALV-SF from a gibbon housed in San Francisco [27] and GALV-Br isolated form gibbon brain material [28].

There have also been two isolates derived as a contaminant from cell cultures. These are GALV-X, found in a HUT-78 cell culture line, which had been infected with HIV [29, 30] and GALV-Mar, which was detected in a cell culture derived from Marmoset cells (unpublished sequence GenBank: U20589.1). There is also a related virus, initially named Simian Sarcoma Associated Virus (SSAV) and now Woolly monkey Virus (WMV). It was isolated from a pet Woolly monkey from California that developed fibrosarcomas and had apparently been housed with a gibbon. Woolly monkey virus is a defective recombinant virus, which has lost its envelope gene and acquired a cellular oncogene [31].

The discovery of a novel oncogenic retrovirus, which was highly pathogenic and which infected sub human primates promoted a great deal of research following its initial discovery. Some researchers believed it might lead to the discovery of a novel human leukaemia virus.

3. GALV-related viruses

Following the discovery of GALV, one of the lines of query focused on what might be the origin of this virus. It had never been isolated before that initial lymphoma outbreak in Bangkok and there are no reports it has ever been seen in wild gibbons. Initially, the focus for the origin of the virus was Southeast Asian rodents because GALV is loosely related to the murine leukaemia viruses. The lack of evidence for prior gibbon infection and the relationship with murine leukaemia viruses suggested that GALV represented a novel cross species transmission event of an unknown virus from a yet to be identified host, possibly a rodent. In the 1970s, several papers were published positing possible candidates for the host of the ancestor virus. These included the Asian rodents *Vandeleuria oleracea, Mus dunni* and *Mus caroli* [32–34]. However, these early papers were based on relatively low specificity technology such as DNA hybridisation, and there is no published retroviral sequence from these rodents that indicate they are the host of a GALV variant.

4. Koala retrovirus

An interesting new chapter on this virus was opened in 2000 when Hanger et al. published the full sequence of a gamma retrovirus isolated from koalas, which was named koala retrovirus (KoRV) [35]. The search for such a virus was prompted by the clinical observation that many wild and captive koalas appeared to be suffering from a high incidence of lymphoid tumours and immunosuppressive like disorders, diseases often associated with retroviral infection [36]. Koala retrovirus is interesting because it appears to be the only known naturally occurring exogenous virus, which is actively undergoing a process of endogenisation in its host [37].

When the full sequence of KoRV was published, it became apparent that it was closely related to GALV, suggesting that the two viruses almost certainly shared a common ancestor [35]. The obvious question from this observation is by what mechanism is a virus able to infect a primate on the Southeast Asian mainland and a marsupial in Australia when the host species are phylogenetically diverse and are separated by thousands of kilometres. It appears there has been some cross species transmission events which are yet to be determined.

5. *Melomys burtoni* retrovirus

Another intriguing aspect was uncovered in 2014 with the publishing of four partial proviral sequences obtained from a native Australian rodent, the grassland melomys, *Melomys burtoni*. This was named *Melomys burtoni* retrovirus (MbRV) and it shares close homology with both GALV and KoRV across the published sequences. A total of 2880 bp were sequenced from the *pol* and *env* genes and they had 94, 93, 92 and 90% nucleotide identity with GALV-SEATO and 84, 82, 74 and 79% identity with KoRV, respectively [38]. *M. burtoni* has a geographic range which in part overlaps that of koalas. It inhabits dry sclerophyll forest similar to that of koala

http://dx.doi.org/10.5772/intechopen.71694

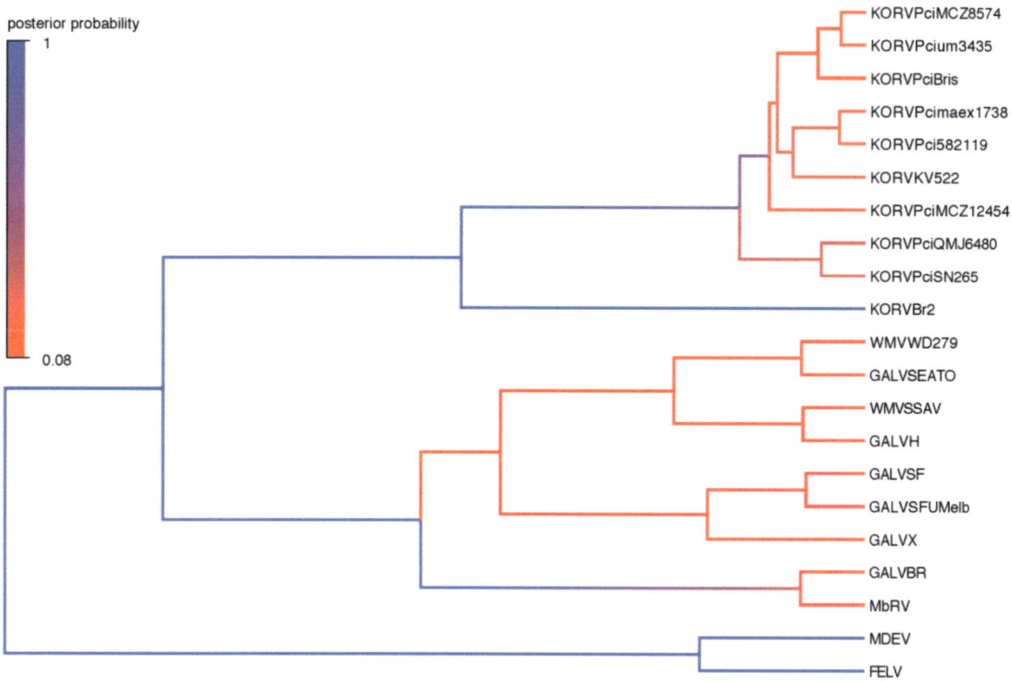

Figure 2. Consensus multi-locus phylogeny showing evolutionary relationships of various gammaretroviruses. Analyses were performed using the *BEAST Bayesian algorithm, which co-estimates multiple gene trees within a shared species phylogenetic matrix. Untrimmed pol (3606 bp) and env (2391 bp) genes were included as separate alignments with unlinked substitution models (HKY + G for pol; GTR + I for env) and evolutionary rates (allowing the sampler to estimate strict clock rates from a uniform (0–1*e100) prior distribution). The two gene partitions were embedded in a species tree matrix using a Yule speciation prior. Two MCMC chains were run for 40,000,000 iterations each, sampling every 2000 iterations (resulting in 30,000 posterior estimates after a 25% burnin). Chains were examined visually to confirm adequate mixing and ensure that estimated parameter sample sizes were above 200. Colours of branches represent posterior probabilities of node placement, with warmer reds showing relatively low support and cooler blues showing high support. GenBank accession numbers for each taxa are as follows: KORVPciMCZ8574 (KF786282); KORVPcium3435 (KF786286); KORVPciBris (AF151794); KORVPcimaex1738 (KF786281); KORVPci582119 (KF786280); KORVKV522 (AB721500); KORVPciMCZ12454 (KF786283); KORVPciQMJ6480 (KF786284); KORVPciSN265 (KF786285); KORVBr2 (KC779547); WMVWD279 (KX059700); GALVSEATO (KT724048); WMVSSAV (KT724051); GALVH (KT724050); GALVSF (KT724047); GALVSFUMelb (X13194); GALVX (U60065); GALVBR (KT724049); MbRV (KF572483–KF572486); MDEV (AF053745) and FELV (NC_001940). (From McKee et al. [40], virus genes).

habitat [39]. Thus, it appears possible that there may have been a direct viral transmission between koalas and *M. burtoni* in the past. This would explain the origin of KoRV in koalas. However, what is intriguing is that while MbRV and KoRV are clearly closely related, MbRV and GALV are closer still. If MbRV had been isolated from a gibbon, it would be listed as another strain of GALV (**Figure 2**).

6. Current theories on the origin of GALV

Recently, there have been three papers published putting forward different scenarios which might explain the origin of GALV.

Brown and Tarlinton suggested that GALV arose iatrogenically by the inoculation of biological material derived from Southeast Asian rodents housed at the SEATO facility [41]. They cite a number of SEATO Medical Research Laboratory Annual Progress Reports from the 1960s and 1970s, which give details of those early experiments. These reports shed some light on possible mechanisms of iatrogenic transmission to gibbons. For example, a colony of the Asian house mouse, *Mus musculus castaneus,* was maintained at the SEATO facility and presumably this colony provided the mice used in experiments there [42]. A colony of laboratory rats presumed to be Wistar rats, which were imported from Malaysia in February 1964, was also maintained [43].

Gibbons were used in many experiments where they were inoculated with biological material including blood and viruses obtained from a range of sources. One SEATO report indicates that gibbons were inoculated with material taken directly from rodents. Mice were used to passage viruses and on one occasion two gibbons were inoculated with ".... a low passage suckling mouse brain (SMB) suspension ..." [44]. While the origin of these mice is not stated, it is reasonable to assume that they came from the *M. musculus castaneus* colony. In addition, one of the prototype DFV strains (type 2 New Guinea C strain) injected into gibbons was repeatedly passaged through suckling mice, at least in the early years [45].

Some research gibbons were kept free ranging on Ko Klet Kaeo, an island just off the coast in the Gulf of Thailand. Rodents known to live on the island were *Rattus rattus* and *Bandicoota indicus* [46]. Since these gibbons were free ranging, it may have been possible for a close interaction allowing a cross-species transmission event. Thus, there were occasions where gibbons were in close proximity with rodents or were injected with biological material acquired from rodents, and this may have allowed the transfer of a GALV progenitor to gibbons. However, there are currently no retroviral sequences from any of the rodent species mentioned above, which indicate they are the host of the ancestral virus.

Bats are known to harbour a large number of endogenous retroviruses and it has been suggested that retroviruses may have first evolved in bats [47, 48]. Denner published an alternative hypothesis suggesting that bats may be the host species for the origins of both GALV and KoRV [49]. Bats, especially fruit bats, unlike most birds and other mammals, freely cross Wallace's line and thus may have carried a virus between Southeast Asia and Australia. Wallace's line is an imaginary line, which passes between the Indonesian islands of Bali and Lombok. It divides the fauna of the region into typically Southeast Asian fauna to the northwest and typically Australasian fauna to the southeast (**Figure 3**). It is named after the nineteenth century zoologist and explorer Alfred Russel Wallace [50]. While there have been some published sequences from bats which are loosely related to GALV, as is the case with rodents from Thailand, there is currently no published sequence which can be considered a GALV ancestor. In addition, there is no single species of bat, which is known to have its geographic distribution extend from Australia to Thailand. Thus, if bats were hosts for a KoRV/GALV ancestor, it would require yet more cross transmission events as more than one species of bat would be needed to account for the spread of a virus between Australia and Thailand.

McKee et al. suggested an alternative theory which provides a possible if unlikely route by which an ancestor virus might have been iatrogenically transmitted to gibbons [40]. Their

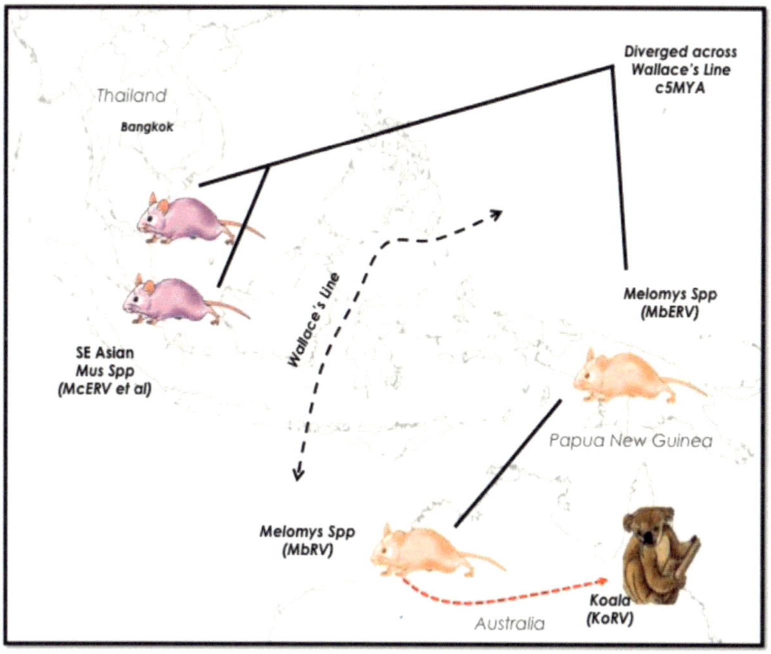

Figure 3. Divergence of Southeast Asian and Australasian rodents: Southeast Asian rodents diverged from the Australasian old endemic rodents around 5 million years ago presumably after being isolated by sea level rise. Wallace's line marks the geographic discontinuity that separates Asian from Australasian fauna. We propose that this also marked the co-divergence in murine ERV hosted by each group with Melomys eventually giving host to MbRV. Some Melomys are semiarboreal raising the possibility of niche overlap and eventual cross-species transmission of MbRV to koalas (*Phascolarctos* spp.). This scenario implies that MbRV is a progenitor murine-like ERV, which stabilised in the Melomys genome but retained infectivity over a few million years. Of note is that MbRV appears to be a complete RV with intact ORFs but it remains to be seen whether it is in fact infective. (From McKee et al. [40], virus genes).

theory is that an infectious MbRV/GALV-related virus is present in melomys species in Papua New Guinea, and further that this virus has been able to infect people or at least was a contaminant when samples were collected from human patients and biologic material derived from these samples later injected into gibbons. They suggest that this happened on at least two separate occasions. In the first, gibbons were inoculated with a New Guinea strain of DFV [51]. Details on how this virus was obtained are not available, but if a dengue fever patient in PNG from whom the New Guinea strain was obtained was concurrently infected with MbRV; this would theoretically have allowed transmission to gibbons. Sometime after this inoculation, the first strain of GALV appeared. No gibbons had been diagnosed with GALV infection prior to these experiments.

In 1968, brain material taken from human kuru patients from the Eastern Highlands of Papua New Guinea was injected into four gibbons at the Gulf South Primate Center, New Iberia, Louisiana, USA. These animals subsequently died of pneumonia and their brains were frozen for later experiments. When this brain material was added to cell cultures, the GALV-Br strain was isolated. It is worth noting that the authors stated "…contact of these gibbons with animals bearing known, experimentally induced infections with primate type C viruses was not possible" [28]. Thus, two strains of GALV were detected in gibbons after injection with biological material obtained from people from Papua New Guinea. An alternative explanation for the appearance of GALV-Br is that it may have arisen as cell culture contaminant in the same way the GALV-X appeared [41].

All three of the above theories are interesting and have positive and negative aspects. A deficiency in the theories that Southeast Asian rodents or bats are the ancestral host is that currently there is no retroviral sequence from any of these species with a sufficient high degree of homology to be considered the ancestor virus. The partial MbRV sequence does have such similarity.

Proponents of the bat and Asian rodent theories for the appearance of GALV appear to be suggesting that a GALV variant exists independently in these species. However, it is difficult to understand how there could be another virus circulating in a host in mainland Southeast Asia while a very similar virus is present in Melomys. *Melomys burtoni* occurs in parts of Australia and Papua New Guinea, and the genus Melomys is restricted to the Australasian side of Wallace's line. It does not occur in Thailand [52]. Melomys are termed "old world endemics." They came down through the land bridge that was present as part of Sahul when Australia and Papua New Guinea were connected approximately 5 million years ago during an adaptive radiation in the Pliocene Epoch [53]. Thus, melomys and mainland Southeast Asian rodents have been isolated for at least 5 million years, and over such a time period, it would be expected that their respective genomes would have diverged significantly given the high mutation rate that exogenous retroviruses undergo [54, 55].

It is possible that bats may have been able to have close interactions with some research gibbons, in particular those present on Ko Klet Kaeo, and as noted above, bats do cross Wallace's line. However, it would seem that these interactions could occur equally with wild gibbons and GALV has only been detected in captive animals.

An issue with the Papua New Guinea origin [40] is that there is currently no evidence that MbRV is infectious. A related virus, designated as Melomys/Woolly monkey virus (MelWMV), detected in a novel *Melomys burtoni* subspecies from Wallacea and clearly related to MbRV is endogenous and has suffered fatal mutations. Thus, it is unlikely to be the origin of GALV. Only partial sequences of MbRV have been published and it may also be a defective endogenous virus incapable of infection. In addition, there is currently no evidence for human infection with an MbRV variant in Papua New Guinea, although it is possible that such an infection may be present as a sub clinical entity.

7. Conclusions

The origins of GALV remain a mystery a half century after it was first detected and it remains a fascinating saga in the field of retrovirology. A number of theories have been proposed which might explain the origins and clearly more research is needed to definitively answer this question. Areas of investigation could include screening of possible vertebrate hosts, such as the Asian rodents that may have had close contact with gibbons at

http://dx.doi.org/10.5772/intechopen.71694

the SEATO facility. These include the *Mus musculus castaneus* and Wistar rat colonies at the facility, and *Rattus rattus* and *Bandicoota indicus* from Ko Klet Kaeo Island. Bats, whose geographic distribution extends to the region around Bangkok, and melomys species from Papua New Guinea could also be examined. In addition, it would be interesting to screen people from Papua New Guinea who live in regions where they may have contact with melomys. Close contact between *Melomys burtoni* and humans is also possible in the coastal regions of northern and eastern Australia where this species occurs, and this again raises the possibility of human infection with this virus. While there is no evidence that any individual working with GALV at the SEATO facility was ever infected, human infection cannot be completely ruled out and screening of people living in close proximity with *Melomys burtoni* would be of interest. In particular, patients suffering from lymphoid tumours could be tested. Lack of appropriate diagnostic facilities in some regions of Papua New Guinea may make this difficult. It should be noted that the ancestor virus may never be found. An infectious exogenous virus circulating in any vertebrate host does not necessarily have a high prevalence of infection. Thus, it is possible that many specimens from the host species could be screened and yield negative results.

Author details

Gregory Stewart Simmons[1]* and Gervais Habarugira[2]

*Address all correspondence to: g.simmons@uq.edu.au

1 School of Veterinary Science, University of Queensland, Gatton, Queensland, Australia

2 School of Animal Sciences and Veterinary Medicine, The University of Rwanda, Kigali, Rwanda

References

[1] Herschhorn A, Hizi A. Retroviral reverse transcriptases. Cellular and Molecular Life Sciences. 2010;**67**:2717-2747

[2] Jern P, Coffin JM. Effects of retroviruses on host genome function. Annual Review of Genetics. 2008;**42**:709-732

[3] Rulli SJ Jr, Hibbert CS, Mirro J, Pederson T, Biswal S, et al. Selective and nonselective packaging of cellular RNAs in retrovirus particles. Journal of Virology. 2007;**81**:6623-6631

[4] Lairmore MD. Retroviridae. In: MacLachlan NJ, Dubovi EJ, editors. Fenners Veterinary Virology. 4th ed. London: Elsevier; 2011. pp. 243-274

[5] Ruprecht K, Mayer J, Sauter M, Roemer K, Mueller-Lantzsch N. Endogenous retroviruses and cancer. Cellular and Molecular Life Sciences. 2008;**65**:3366-3382

[6] Stoye JP. Endogenous retroviruses: Still active after all these years? Current Biology. 2001;**11**:R914-R916

[7] Hayward A, Katzourakis A. Endogenous retroviruses. Current Biology. 2015;**25**:R644-R646

[8] Gifford R, Tristem M. The evolution, distribution and diversity of endogenous retroviruses. Virus Genes. 2003;**26**:291-315

[9] Stocking C, Kozak CA. Murine endogenous retroviruses. Cellular and Molecular Life Sciences. 2008;**65**:3383-3398

[10] Johnson WE. Endogenous retroviruses in the genomics era. Annual Review of Virology. 2015;**2**:135-159

[11] Britannica E, King T. Encyclopaedia Britannica Almanac 2009. Encyclopaedia Britannica, New York; 2009

[12] Sverdlov ED. Retroviruses, their domesticated relatives and other retroinvaders. In: Sverdlov ED, editor. Retroviruses and Primate Genome Evolution. Georgetown, Texas: Eureka.com/Landes Bioscience; 2005. pp. 93-103

[13] Moore PS, Chang Y. Why do viruses cause cancer? Highlights of the first century of human tumour virology. Nature Reviews. Cancer. 2010;**10**:878-889

[14] Magden E, Quackenbush SL, VandeWoude S. FIV associated neoplasms–A mini-review. Veterinary Immunology and Immunopathology. 2011;**143**:227-234

[15] Hartmann K. Clinical aspects of feline retroviruses: A review. Virus. 2012;**4**:2684-2710

[16] Roszkiewicz J, Smolewska E. Kaleidoscope of autoimmune diseases in HIV infection. Rheumatology International. 2016;**36**:1481-1491

[17] Brockelman WY, Ross BA, Pantuwatana S. Social correlates of reproductive success in the gibbon colony on Ko Klet Kaeo, Thailand. American Journal of Physical Anthropology. 1973;**38**:637-640

[18] Halstead SB. Epidemiological studies of Thai haemorrhagic fever, 1962-64. Bulletin of the World Health Organization. 1966;**35**:80-81

[19] Morris JH. The Ecology of the Gibbon (*Hylobates lar* lar) in: SEATO Medical Research Laboratory Annual Progress Reports. Bangkok, Thailnad: Armed Forces Reseach Institute of Medical Science; 1966

[20] Johnsen DO, Tanticharoenyos P, Pulliam JD. Gibbon Leukaemia. Bangkok, Thailand: US Army-SEATO Medical Research Institute; 1969. pp. 196-197

[21] Johnsen DO, Wooding WL, Tanticharoenyos P, Bourgeois CH. Malignant lymphoma in the gibbon. Journal of the American Veterinary Medical Association. 1971;**159**:563-566

[22] Newberne TG, Robinson VB. Spontaneous tumours in primates—A report of two cases with notes on the apparent low incidence of neoplasms in sub human primates. American Journal of Veterinary Research. 1960;**21**:150-155

[23] DiGiacomo RF. Burkitt's lymphoma in a white-handed gibbon (*Hylobates lar*). Cancer Research. 1967;**27**:1178-1179

[24] Kawakami T, Buckley PM, McDowell TS, DePaoli A. Antibodies to simian C-type virus antigen in sera of gibbons (*Hylobates* sp.). Nature: New Biology. 1973;**246**:105-107

[25] Alfano N, Kolokotronis SO, Tsangaras K, Roca AL, Xu W, et al. Episodic diversifying selection shaped the genomes of gibbon ape leukemia virus and related gammaretroviruses. Journal of Virology. 2015;**90**:1757-1772

[26] Krakower JM, Tronick SR, Gallagher RE, Gallo RC, Aaronson SA. Antigenic characterization of a new gibbon ape leukemia virus isolate: Seroepidemiologic assessment of an outbreak of gibbon leukemia. International Journal of Cancer. 1978;**22**:715-720

[27] Kawakami TG, Huff SD, Buckley PM, Dungworth DL, Synder SP, et al. C-type virus associated with gibbon lymphosarcoma. Nature: New Biology. 1972;**235**:170-171

[28] Todaro GJ, Lieber MM, Benveniste RE, Sherr CJ. Infectious primate type C viruses: Three isolates belonging to a new subgroup from the brains of normal gibbons. Virology. 1975;**67**:335-343

[29] Burtonboy G, Delferriere N, Mousset B, Heusterspreute M. Isolation of a C-type retrovirus from an HIV infected cell line. Archives of Virology. 1993;**130**:289-300

[30] Parent I, Qin Y, Vandenbroucke AT, Walon C, Delferriere N, et al. Characterization of a C-type retrovirus isolated from an HIV infected cell line: Complete nucleotide sequence. Archives of Virology. 1998;**143**:1077-1092

[31] Theilen GH, Gould D, Fowler M, Dungworth DL. C-type virus in tumor tissue of a woolly monkey (*Lagothrix* spp.) with fibrosarcoma. Journal of the National Cancer Institute. 1971;**47**:881-889

[32] Callahan R, Meade C, Todaro GJ. Isolation of an endogenous type C virus related to the infectious primate type C viruses from the Asian rodent *Vandeleuria oleracea*. Journal of Virology. 1979;**30**:124-131

[33] Lieber MM, Sherr CJ, Todaro GJ, Benveniste RE, Callahan R, et al. Isolation from the asian mouse *Mus caroli* of an endogenous type C virus related to infectious primate type C viruses. Proceedings of the National Academy of Sciences of the United States of America. 1975;**72**:2315-2319

[34] Wolgamot G, Bonham L, Miller AD. Sequence analysis of *Mus dunni* endogenous virus reveals a hybrid VL30/gibbon ape leukemia virus-like structure and a distinct envelope. Journal of Virology. 1998;**72**:7459-7466

[35] Hanger JJ, Bromham LD, McKee JJ, O'Brien TM, Robinson WF. The nucleotide sequence of koala (*Phascolarctos cinereus*) retrovirus: A novel type C endogenous virus related to gibbon ape leukemia virus. Journal of Virology. 2000;**74**:4264-4272

[36] Canfield PJ, Sabine JM, Love DN. Virus particles associated with leukaemia in a koala. Australian Veterinary Journal. 1988;**65**:327-328

[37] Tarlinton RE, Meers J, Young PR. Retroviral invasion of the koala genome. Nature. 2006;**442**:79-81

[38] Simmons GS, Clarke D, McKee JJ, Young P, Meers J. Discovery of a novel retrovirus sequence in an Australian native rodent (*Melomys burtoni*): a putative link between gibbon ape leukemia virus and koala retrovirus. PLOS One. 2014 Sep 24;**9**(9):e106954

[39] Redhead T. Mosaic tailed rats. In: Strahan R, editor. Complete Book of Australain Mammals. Angus and Roberston: Melbourne; 1983. pp. 370-379

[40] McKee JJ, Clark N, Shapter F, Simmons G. A new look at the origins of gibbon ape leukemia virus. Virus Genes 2017;**53**:165-172

[41] Brown K, Tarlinton R. Is gibbon ape leukaemia virus still a threat. Mammal Review. 2016;**47**:53-61

[42] Marshall JT. Vertebrate reservoirs of disease. In: SEATO Medical Research Laboratory Annual Progress Reports AFRIMS; 1975

[43] Morris JH. Nutritional and health requirements for development and maintainance of conventional animal colonies. In: SEATO Medical Research Laboratory Annual Progress Reports; 1965

[44] Bancroft WH, Snitban R, Rozmiarek H, Tingpalapong M. Experimental infection of gibbons with a Group B Arbovirus (T-1674). In: SEATO Medical Research Laboratory Annual Progress Reports; 1975

[45] Sinarachatanant P, Olson LC. Replication of dengue virus type 2 in *Aedes albopictus* cell culture. Journal of Virology. 1973;**12**:275-283

[46] Berkson G. Ko Klet Kaeo: Habitat description. In: SEATO Medical Research Laboratory Annual Progress Reports; 1968

[47] Cui J, Tachedjian G, Tachedjian M, Holmes EC, Zhang S, et al. Identification of diverse groups of endogenous gammaretroviruses in mega- and microbats. The Journal of General Virology. 2012;**93**:2037-2045

[48] Cui J, Tachedjian M, Wang L, Tachedjian G, Wang LF, et al. Discovery of retroviral homologs in bats: Implications for the origin of mammalian gammaretroviruses. Journal of Virology. 2012;**86**:4288-4293

[49] Denner J. Transspecies transmission of gammaretroviruses and the origin of the gibbon ape leukaemia virus (GaLV) and the koala retrovirus (KoRV). Virus. 2016;**8**(12). pii: E336. DOI: 10.3390/v8120336

[50] Bastin J. Introduction. The Malay Archipelego: Oxford University Press; 1989. pp. vii-xxvii

[51] Nisalak A, Gunakasem P, Russell P, Singharaj P, Udomsakdi S. Arbovirus infections in man and experimental animals. In: SEATO Medical Research Laboratory Annual Progress Reports; 1966

[52] Nowak R. Mosaic tailed rats. In: Nowak R, editor. Walkers Mammals of the World. Baltimore and London: Johns Hopkins University Press; 1999. pp. 1548-1550

[53] Rowe KC, Reno ML, Richmond DM, Adkins RM, Steppan SJ. Pliocene colonization and adaptive radiations in Australia and new Guinea (Sahul): Multilocus systematics of the old endemic rodents (Muroidea: Murinae). Molecular Phylogenetics and Evolution. 2008;**47**:84-101

[54] Overbaugh J, Bangham CRM. Selection forces and constraints on retroviral sequence variation. Science (New York, N.Y.). 2001;**292**:1106-1109

[55] Andreoletti L, Reveil B, Moret H, Brodard V, Philbert F, et al. Significant genetic and antigenic variability within the env gene of systemic human immunodeficiency virus type 1 group O populations during the natural course of a heterosexual infection: A pilot study. Journal of Clinical Microbiology. 2007;**45**:1319-1321

Chapter 8

Trypanosoma cruzi Infection in Non-Human Primates

Renato Sathler-Avelar,
Armanda Moreira Mattoso-Barbosa,
Olindo Assis Martins-Filho,
Andrea Teixeira-Carvalho,
Danielle Marchetti Vitelli-Avelar,
John L. VandeBerg and Jane F. VandeBerg

Additional information is available at the end of the chapter

http://dx.doi.org/10.5772/intechopen.71652

Abstract

For decades, non-human primates (NHPs) have been employed as experimental models to study many aspects of human diseases. They are the closest genetically to humans of any of the models applied in biomedical research; therefore, many authors have published scientific work regarding these animals and infectious diseases, including tuberculosis, AIDS, and tropical diseases. Among these, Chagas disease has caught the attention of many researchers all over the world. Recent studies have demonstrated great similarities with the human pathology, including cardiomyopathy and exacerbated pro-inflammatory response. Besides being genetically close to humans, NHP have a great probability to be naturally infected by *Trypanosoma cruzi*, which turns them into more interesting models to study Chagas disease mechanisms.

Keywords: non-human primates, Chagas disease, *T. cruzi*, immunology, infectious diseases

1. Introduction

The haemoflagellate *Trypanosoma cruzi* causes Chagas disease, one of the most relevant neglected tropical diseases of humankind. The World Health Organization estimates that there are 6–7 million people infected over the world [1–3] as shown in **Figure 1**. Nevertheless, other mammals are also at risk of becoming infected, such as marsupials, armadillos, sylvatic and domestic dogs, racoons and non-human primates (NHPs) [4–8]. The most common Chagas disease transmission is the vectorial via by several species of triatomine. The

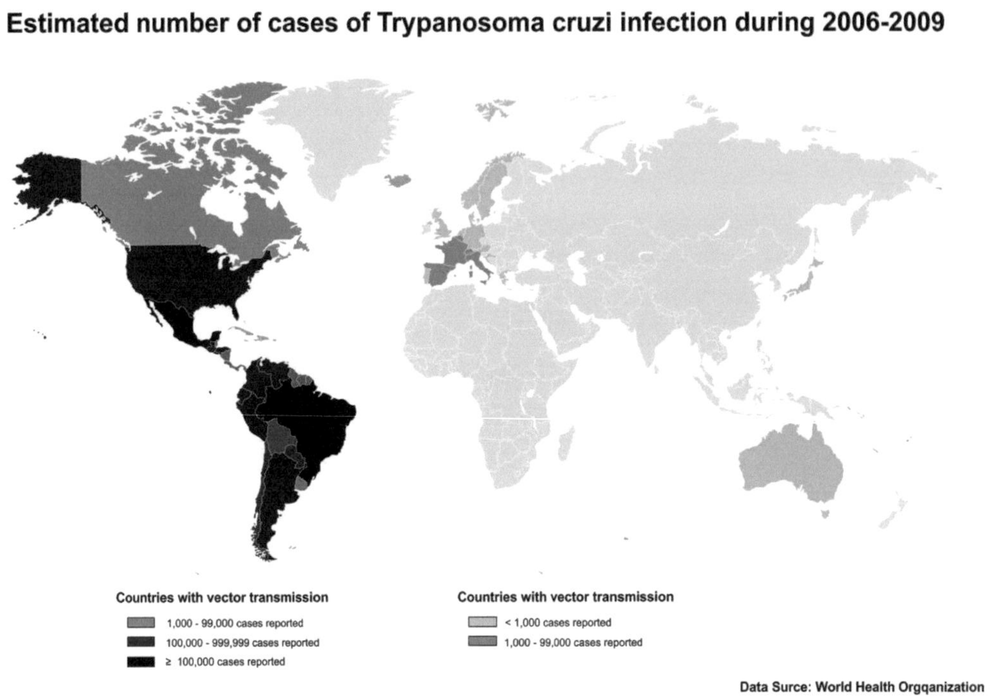

Figure 1. Descriptive map of infection areas in the world.

vector insect ingests a blood meal containing bloodstream trypomastigotes which later, in the insect's gut, the parasite differentiates into epimastigotes and replicates. When the vector defecates, metacyclic trypomastigote forms are released and invade the host through broken skin or mucosal membranes. A brief schematic mechanism of *T. cruzi* infection is shown

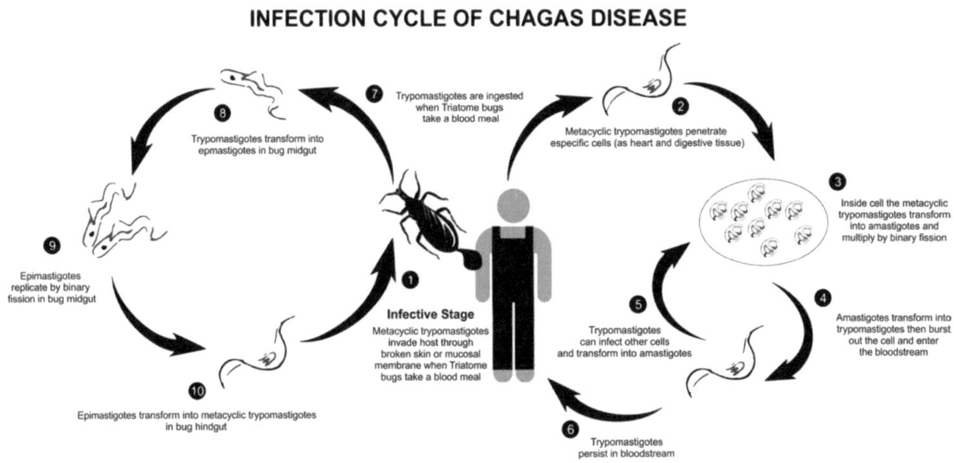

Figure 2. Schematic life cycle of *Trypanosoma cruzi.*

in **Figure 2**. In addition, the oral and congenital transmissions are other important ways of becoming infected [6]. Out of these, the oral transmission seems more relevant, especially by non-human species. The habit of consuming bugs may predispose these animals to ingest infected triatomines [9]. Amongst the non-human species presented above, non-human primates are the greatest species comparable to human beings, leading researchers all over the world to employ NHP in biomedical-related studies. Interestingly, Carlos Chagas was the first to describe both experimental and natural *T. cruzi* infections in non-human primates [10]. After that, many others have portrayed the disease in these animals, and studies are still being produced nowadays. From small college laboratories to huge pharmaceutical industries, the purpose is the same: to find better options to diagnose and to treat patients. Our group has been working with NHP for a few years, and so far, our findings are similar to those from many researchers all over the world. The aim of this work is to explore the most relevant findings regarding *T. cruzi*-infected NHP.

2. Clinical manifestations of NHP *T. cruzi* infection

Several studies have demonstrated that NHP develop clinical manifestations highly similar to what is observed in both acute and chronic human Chagas disease [9, 11–14]. In the acute phase, several signs and symptoms can be observed, such as inoculation chagoma, patent parasitemia, *T. cruzi*-specific IgM and IgG antibodies, and leukocytosis and lymphocytosis. Histopathological data revealed intense heart parasitism and pronounced inflammatory infiltrate, along with myocardial fibrosis with collagen deposits [12]. Besides that, cardiac alterations have also been found in these animals, such as abnormal electrocardiogram and heart muscle cells presenting degrees of damage [13]. Those findings reinforce the results found by Bonecini-Almeida et al. [11] who described electrocardiographic patterns detected in *T. cruzi*-infected rhesus monkeys during the acute phase. The results have evidenced atrio-ventricular block, right bundle branch block—first-degree His bundle, low voltage QRS complex, and abnormal ventricular repolarization. Interestingly, these alterations disappeared at the fourth month post infection. Controversially, Bommineni et al. [14] demonstrated that the acute phase in NHP may be lethal. Despite that, *T. cruzi* infection usually evolves from an acute phase to a chronic phase that may manifest itself in a variety of ways.

In contrast to the acute phase, during the chronic stages of the disease, the trypomastigotes in the peripheral circulation are extremely difficult to detect microscopically. However, more detailed histopathological studies have shown nests of *T. cruzi* amastigote forms in host cardiac tissue from naturally infected baboons. The majority of individuals that progress to the chronic phase remain clinically asymptomatic for many years, characterizing the indeterminate clinical form of the disease. Usually the disease confirmation requires the application of several diagnostic techniques, such as microscopic examination of blood smears, serological assays, xenodiagnosis, hemoculture and PCR-based assays for direct detection and quantification of parasite DNA [15–17]. After long years of infection, individuals may progress to the cardiac and/or digestive chronic phase, which usually represents the most severe clinical damages [18]. Researchers have demonstrated numerous alterations in the electrical conduction system, ventricular arrhythmias, cardiomegaly, enteromegaly, myocardial fibrosis, and edema, along with other clinical signs of chronicity [9, 12, 13, 19, 20]. In several cases the infected NHP develop aggressive chronic Chagas disease and die, usually due to cardiac damages. In this context, a NHP animal model with these features could contribute significantly to better comprehend the disease outcome as well as to improve the therapies for Chagas disease.

3. Immunological features of NHP *T. cruzi* infection

In addition to pathophysiological changes, NHP also manifest alterations in their immunological system. It is well known that the immune system plays an important role in the pathogenesis of the disease. When it comes to NHP, it is not different. Recent studies, including ours, have shown that their immunological response resembles what is observed in human Chagas disease, as briefly schemed in **Figure 3** [1, 6, 18, 21].

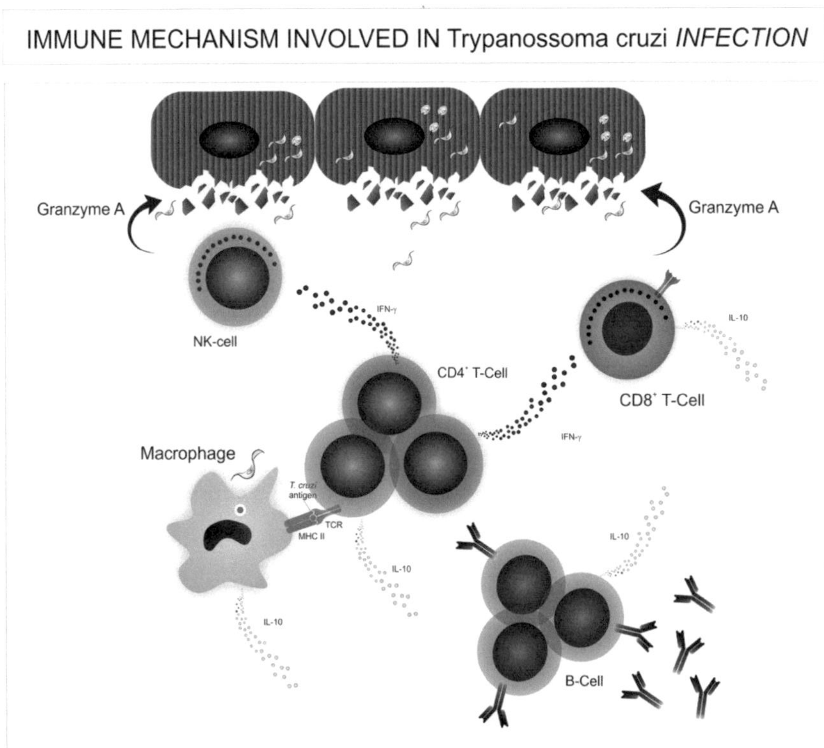

Figure 3. *Trypanosoma cruzi* naturally infected non-human primate displays, in the peripheral blood, a high activity of cytotoxic cells (Granzyme A$^+$ NK cells and Granzyme A$^+$/Perforin CD8$^+$ T cells) and expansion of macrophages and activated T-cell subsets. Furthermore, the infected animals exhibit an overall mixed pro-inflammatory/regulatory cytokine milieu, with CD4$^+$ T cells, the most important source of IFN-γ, as well as CD4$^+$ T cells, CD8$^+$ T cells, macrophage/monocytes and B-cell producers of IL-10.

Our group has recently published a research on *T. cruzi* naturally infected cynomolgus macaques which displayed, in the peripheral blood, a similar immunological profile to that observed in humans, with high activity of cytotoxic cells and expansion of macrophages and activated T-cell subsets [21]. The infected animals exhibit higher frequency of NK Granzyme A^+ cells. Furthermore, this cell population was able to increase the pro-inflammatory cytokine secretion afterwards *T. cruzi* antigen stimulation [22]. These data reinforce the important role of NK cells as a source of IFN-γ to activate macrophages and increase the nitric oxide production to inhibit the intracellular parasite growth [23, 24]. Moreover, the NK cells mediate a relevant cytotoxic mechanism that kill infected host cells or even free parasites throughout a lytic perforin-independent mechanism [24]. It is important to mention that the higher expression of inducible nitric oxide synthase by monocytes/macrophage has been correlated with loss of connexin43 in cardiopathic *T. cruzi*-infected rhesus monkeys [12]. Connexin43 is the major protein responsible for the electrical synchrony of cardiomyocytes [25]. In this context, any injury in this protein may result in arrhythmias and heart failure during the chronic chagasic cardiomyopathy.

It is well known that the adaptive immune response plays a critical role in Chagas disease progression in humans; however, in NHP its mechanisms remain unclear. Recently, our group showed that the *T. cruzi*-infected NHP developed a pattern of activated T lymphocytes as observed in the human infection. In fact, higher expression of CD54 and HLA-DR by T cells, especially within the $CD8^+$ subset, along with outstanding expression of Granzyme A and Perforin, emphasized the enhanced cytotoxicity-linked pattern of $CD8^+$ T lymphocytes. These data reinforce the role of $CD8^+$ T lymphocytes in the pathogenesis of Chagas disease. Additionally, Pisharath et al. [6] while evaluating *T. cruzi* naturally infected cynomolgus macaques demonstrated by immunohistochemistry that the inflammatory infiltrate from cardiac tissue had mild to moderate multifocal areas, composed predominantly by $CD8^+$ T cells and $CD68^+$ monocyte/macrophage with fewer $CD4^+$ T lymphocytes. In agreement with these data, Mubiru et al. [20] showed a focal and multifocal collection of lymphocytes and plasma cells, as well as rare granulocyte infiltration within the myocardium and epicardium. Moreover, their study revealed a positive correlation between PCR positivity and lymphocytic myocarditis in both baboons and cynomolgus macaques infected with *T. cruzi*, reinforcing the hypothesis of direct parasite-induced damage and *T. cruzi*-specific immune responses, in myocardial injury.

It is known that B lymphocytes play a crucial role in protecting against *T. cruzi*. This is due to the fact that these cells synthesize anti-*T. cruzi* antibodies, establish the functional pattern of T-cell cytokines and still are involved in the maintenance of $CD8^+$ memory cells [26, 27]. In addition, it has been displayed that NHP infected with *T. cruzi* presents a high frequency of B-cell population associated with upregulated expression of Fc-γRII (CD32), enhancing the potential of this biomarkers' high expression, in counterbalancing the $CD8^+$ T-cell cytotoxic activity and influencing the degree of myocardiopathy.

It has been clear that cytokines are integral components of the complex intercellular system required to mount and control disease morbidity [28, 29]. However, little is known about the cytokine profile during NHP infection with *T. cruzi*. In order to further understand the mechanisms of *T. cruzi* infection in NHP, Vitelli-Avelar [22], for the first time, characterized the ex vivo cytokine pattern of cynomolgus macaques naturally infected with *T. cruzi* and observed an overall mixed pro-inflammatory/regulatory cytokine milieu, mediated by IFN-γ from $CD4^+$ T cells counterbalanced by IL-10 produced by $CD4^+$ T cells and B cells. This microenvironment resembles that previously described for chronic Chagas disease in humans, mainly in indeterminate clinical form [24]. It has been proposed that this pro-inflammatory/regulatory pattern represents a key element to control deleterious antiparasite immune-mediated inflammatory mechanisms [30].

T. cruzi strains are currently classified into six discrete typing units (DTUs) named TcI to TcVI. It is known that these DTUs have different biological and geographical features [31]. In South American isolates, all of the strains have been characterized from a variety of host species. In contrast, isolates from the Central and North America have been characterized only as TcI or TcIV [32]. Several researchers have discussed the characteristics of different *T. cruzi* genotypes, and it seems like that the strain diversity is associated with the distinct immunological patterns observed in Chagas disease, which might be associated to disease severity [31, 33–36]. While working with a North American NHP colony, we intended to provide insights pertinent to the higher prevalence of TcI natural infection observed amongst these animals by interpreting the differential impact of TcI and TcIV antigen priming in vitro on circulating leukocytes. In this context, our data showed that NHP presents distinct cytokine profile in the presence of TcI and TcIV antigen. While the TcIV antigen triggered an outstanding response, characterized by high levels of TNF- and IFN-γ-producing $CD8^+$ T cells, along with low levels of IL-10, the TcI antigen elicits a predominant regulatory microenvironment, mediated by IL-10 derived from HLA-DR^{++} monocytes and T cells with low levels of TNF^+CD8^+ T cells [22]. The prominent pro-inflammatory milieu, mediated by TNF, seems to be relevant to control the *T. cruzi* infection NHP. The role of TNF in protective mechanisms has been already reported, underscoring its ability to activate macrophages and induce nitric oxide production [37]. Additionally, the enhanced frequency of IFN-γ^+ T cells beside low levels of IL-10-producing cells may also account as a relevant trypanocidal event favoring the TcIV clearance. Conversely, the IL-10-mediated microenvironment observed upon TcI-antigenic recall in vitro represents a critical event to support the ongoing infection with the TcI genotype. These findings may support, at least in part, the predominance of TcI infection amongst cynomolgus macaques in Southern part of the United States. **Figures 4** and **5** present a synthesized scheme of our newest cytokine findings and reinforce the distinct cytokine pattern produced upon *T. cruzi* TcI and TcIV antigen recall in vitro [22]. Furthermore, other studies have confirmed the presence of both TcI and TcIV isolates from Amazonian primates, TcI being more predominate strain than TcIV [38–40].

Regardless the relevance of therapeutic intervention to control morbidity and clinical progression of Chagas disease, currently, there are only two drugs available to treat infected hosts, benznidazole and nifurtimox. Several studies have demonstrated that the effectiveness of therapeutic agents against *T. cruzi* is influenced by the parasite load, genotype as well as by intrinsic features of the host immune response. Studies focusing on aspects related to the synergic effect of the immune response and chemotherapeutic agents in humans and NPH are still scarce. Sathler-Avelar and colleagues [21, 22, 24] have provided insights about

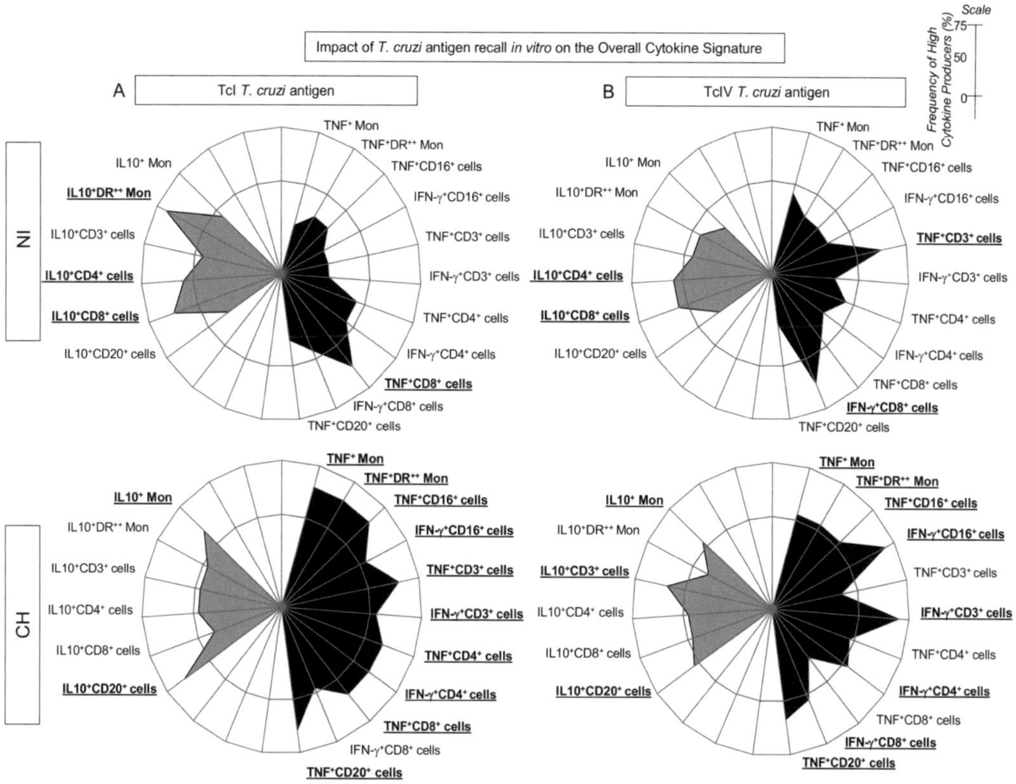

Figure 4. The cytokine milieu in *T. cruzi*-infected and non-infected non-human primates upon antigen recall from TcI and TcIV *T. cruzi* strain. The radar charts illustrate the changes on the pro-inflammatory (black background) and regulatory cytokine microenvironment (gray background) connecting circulating leukocytes of *T. cruzi*-infected cynomolgus macaques (CH) and non-infected controls (NI) upon TcI (A) and TcIV (B) *T. cruzi* antigenic recall in vitro. Relevant data comprising biomarkers with frequency of producers above the 50th percentile are underscored by bold/underlined font.

the relevance of a balanced immune response elicited after chemotherapeutic intervention to mediated parasite killing but minimize tissue damage. There are evidences supporting that a pro-inflammatory response mediated by IFN-γ acts synergistically with the drug treatment to accomplish effective trypanocidal events [24, 41] and that simultaneous regulatory mechanisms elicited by IL-10 are relevant to control deleterious effects of therapeutic intervention [21, 22, 24].

The urgent need of novel drugs to treat Chagas disease has stimulated scientific community to validate appropriate experimental model or in vitro tolls to conducted studies during preclinical trials. These studies that can contribute and elucidate drug mechanisms are still unknown, in an attempt to find a more effective therapeutic agent. In this context, NHP models have been considered one of the most appropriate tolls, especially due to the similarities between the disease aspects and the immune response observed in NPH as compared to humans. Vitelli-Avelar and colleagues have recently provided data focusing on the immune response of NPH infected with *T. cruzi* that can be used to shed light on this issue. Using an in vitro system of antigen recall to mimicry the endogenous booster of parasite-derived antigens that

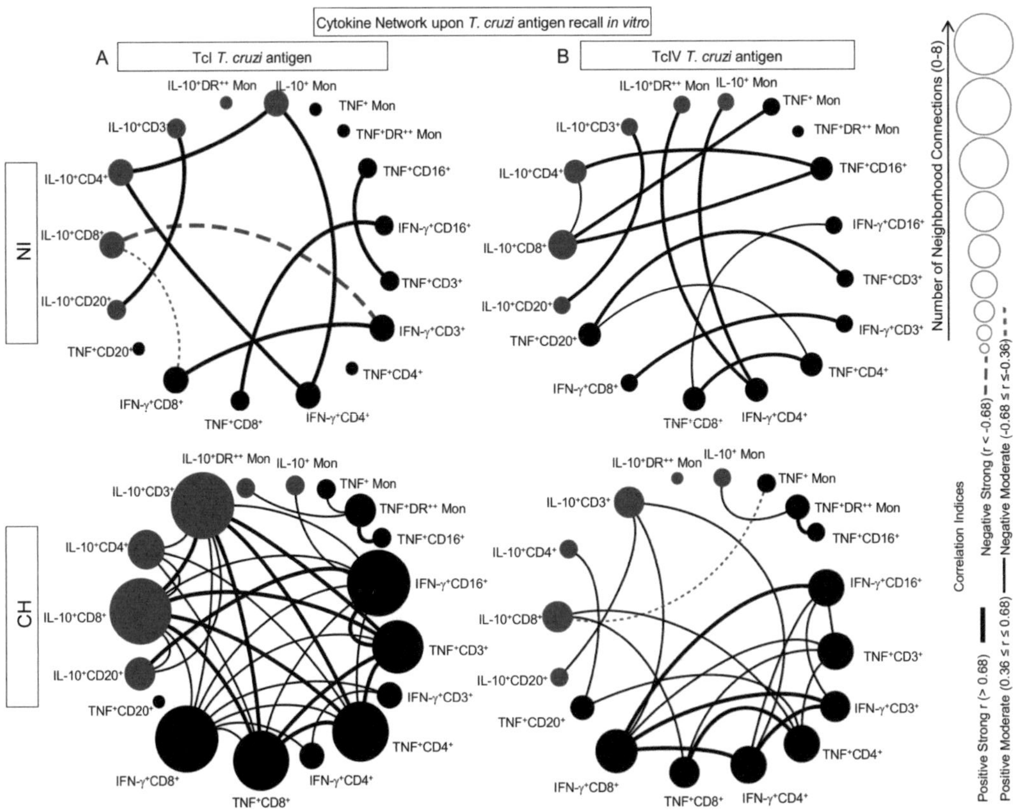

Figure 5. Cytokine network analysis upon *T. cruzi* antigen (TcI/TcIV) stimulation in vitro. Correlation matrices for cytokine producers were constructed to illustrate the distinct cytokine pattern upon (A) TcI and (B) TcIV antigen stimulation. Cytokine$^+$ cell networks for non-human primate naturally infected with *T. cruzi* (CH) and control animals (NI) are shown by clustered distribution of nodes for pro-inflammatory (black) and modulatory (gray) cytokine patterns.

occur throughout chronic infection or upon the extensive antigen release mediated by therapeutic intervention, these authors have demonstrated that similarly to what was found in human Chagas disease patients, NPH-infected host also exhibited a pro-inflammatory/regulatory cytokine signature triggered by *T. cruzi*-antigenic restimulation in vitro, These findings suggest the ability of these hosts to mount an appropriate immune response with putative balanced profile that may contribute for parasite killing, by IFN-γ release, modulated by IL-10 to prevent deleterious idiosyncrasy.

4. Final remarks

The urge for an experimental model that resembles all medical disorders observed in humans is of great importance. Non-human primates are great models to study Chagas disease. It is clear that these mammals present clinical, immunological, and histopathological resemblances to humans. All studies conducted so far lead to believe that as shown in humans, primates naturally infected with *T. cruzi* also evolve to chronic phase and that is probably associated to the extension of the immune response they develop. With an experimental model that develops clinical and immunological manifestations closely comparable to humans, innovative therapeutic strategies may be deeper studied and new drugs may be developed. There are still much more to comprehend; however, the scientific advances and better comprehension of the mechanisms of *T. cruzi* infection may contribute to find hope to Chagas disease patients.

Author details

Renato Sathler-Avelar[1]*, Armanda Moreira Mattoso-Barbosa[1], Olindo Assis Martins-Filho[2], Andrea Teixeira-Carvalho[2], Danielle Marchetti Vitelli-Avelar[2], John L. VandeBerg[3,4] and Jane F. VandeBerg[3]

*Address all correspondence to: renatosavelar@gmail.com

1 Faculdade de Minas – FAMINAS-BH, Belo Horizonte, MG, Brazil

2 Grupo Integrado de Pesquisas em Biomarcadores, Centro de Pesquisas Rene Rachou, FIOCRUZ, Belo Horizonte, MG, Brazil

3 Southwest National Primate Research Center, Texas Biomedical Research Institute, San Antonio, TX, USA

4 South Texas Diabetes and Obesity Institute, School of Medicine, The University of Texas Rio Grande Valley, Brownsville/Harlingen/Edinburg, Texas, USA

References

[1] Nagajyothi F, Machado FS, Burleigh BA, Jelicks LA, Scherer PE, Mukherjee S, et al. Mechanisms of *Trypanosoma cruzi* persistence in Chagas disease. Cellular Microbiology. 2012;**14**(5):634-643

[2] Teixeira ARL, Hecht MM, Guimaro MC, Sousa AO, Nitz N. Pathogenesis of Chagas' disease: Parasite persistence and autoimmunity. Clinical Microbiology Reviews. 2011; **24**(3):592-630

[3] WHO. World Health Organization [Internet]. 2016 [Updated: 2016]. Available from: http://www.who.int/mediacentre/factsheets/fs340/en/ [Accessed: 09-02-2017]

[4] Tenney TD, Curtis-Robles R, Snowden KF, Hamer SA. Shelter Dogs as sentinels for *Trypanosoma cruzi* transmission across Texas, USA. Emerging Infectious Diseases. 2014; **20**(8):1323-1326

[5] Roellig DM, McMillan K, Ellis AE, Vandeberg JL, Champagne DE, Yabsley MJ. Experimental infection of two South American reservoirs with four distinct strains of *Trypanosoma cruzi*. Parasitology. 2010;**137**(6):959-966

[6] Pisharath H, Zao CL, Kreeger J, Portugal S, Kawabe T, Tarea Burton T, et al. Immunopathologic characterization of naturally acquired *Trypanosoma cruzi* infection and cardiac sequalae in cynomolgus macaques (*Macaca fascicularis*). Journal of the American Association for Laboratory Animal Science. 2013;**52**(5):545-552

[7] Dorn PL, Perniciaro L, Yabsley MJ, Roellig DM, Balsamo G, Diaz J, et al. Autochthonous transmission of *Trypanosoma cruzi* Louisiana. Emerging Infectious Diseases. 2007;**13**(4): 605-607

[8] Dorn PL, Daigle ME, Combe CL, Tate AH, Stevens L, Phillippi-Falkenstein KM. Low prevalence of Chagas parasite infection in a nonhuman primate colony in Louisiana. Journal of the American Association for Laboratory Animal Science. 2012;**51**(4):443-447

[9] Zabalgoitia M, Ventura J, Anderson L, Kd C, Jt W, Jl V. Morphologic and functional characterization of chagasic heart disease in non-human primates. The American Journal of Tropical Medicine and Hygiene. 2003;**68**(2):248-252

[10] Minuzzi-Souza TTC, Nitz N, Knox MB, Reis F, Hagström L, Cuba CAC, et al. Vector-borne transmission of *Trypanosoma cruzi* among captive Neotropical primates in a Brazilian zoo. Parasites & Vectors. 2016;**9**(39):1-6. DOI: 10.1186/s13071-016-1334-7

[11] Bonecini-Almeida MG, Galvão-Castro B, Pessoa MHR, Pirmez C, Laranja F. Experimental Chagas Disease in Rhesus Monkeys. I. Clinical, Parasitological, Hematological and Anatomo-Pathological Studies in the Acute and Indeterminate Phase of the Disease. Memórias do Instituto Oswaldo Cruz. 1990;**85**(2):163-171

[12] Carvalho CME, Silverio JC, Silva AA, Pereira IR, Coelho JMC, Britto CA, et al. Inducible Nitric Oxide Synthase in Heart Tissue and Nitric Oxide in Serum of *Trypanosoma cruzi*-Infected Rhesus Monkeys: Association with Heart Injury. PLoS Neglected Tropical Diseases. 2012;**6**(5):e1644. DOI: 10.1371/journal.pntd.0001644

[13] Carvalho CME, Andrade MCR, Xavier SS, Mangia RHR, Britto CC, Jansen AM, et al. Chronic Chagas' Disease In Rhesus Monkeys (*Macaca Mulatta*): Evaluation of Parasitemia, Serology, Electrocardiography, Echocardiography, and Radiology. The American Journal of Tropical Medicine and Hygiene. 2003;**68**(6):683-691

[14] Bommineni YR, Dick Jr. EJ, Estep JS, Van de Berg JL, Hubbard GB. Fatal acute chagas disease in a Chimpanzee. Journal of Medical Primatology. 2009;**38**(4):247-251. DOI: 10.1111/j.1600-0684.2009.00348.x

[15] Andrade MCR, Dick EJ Jr, Guardado-Mendoza R, Hohmann ML, Mejido DCP, VandeBerg JL, et al. Nonspecific lymphocytic myocarditis in baboons is associated with *Trypanosoma cruzi* Infection. The American Journal of Tropical Medicine and Hygiene. 2009;**81**(2):235-239

[16] Williams JT, Dick EJ Jr, VandeBerg JL, Hubbard GB. Natural chagas disease in four baboons. Journal of Medical Primatology. 2009;**38**(2):107-113. DOI: 10.1111/j.1600-0684.2008.00308.x

[17] Williams JT, Mubiru JN, Schlabritz-Loutsevitch NE, Rubicz RC, VandeBerg JL, Dick Jr EJ, et al. Polymerase chain reaction detection of *Trypanosoma cruzi* in *Macaca fascicularis* using archived tissues. The American Journal of Tropical Medicine and Hygiene. 2009;**81**(2):228-234

[18] Dickerson MF, Astorga NG, Astorga NR, Lewis AD. Chagas disease in 2 geriatric rhesus macaques (*Macaca mulatta*) housed in the Pacific Northwest. Comparative Medicine. 2014;**64**(4):323-328

[19] Zabalgoitia M, Ventura J, Anderson L, Williams JT, Carey KD, VandBerg JL. Electrocardiographic findings in naturally acquired chagasic heart disease in nonhuman primates. Journal of Electrocardiology. 2003;**36**(2):155-160

[20] Mubiru JN, Yang A, Dick Jr EJ, Owston M, Sharp RM, VandeBerg JF, et al. Correlation between presence of *Trypanosoma cruzi* DNA in heart tissue of baboons and cynomolgus monkeys, and lymphocytic myocarditis. The American Journal of Tropical Medicine and Hygiene. 2014;**90**(4):627-633

[21] Sathler-Avelar R, Vitelli-Avelar DM, Mattoso-Barbosa AM, Perdigão-de-Oliveira M, Costa RP, Elói-Santos SM, et al. Phenotypic features of circulating leukocytes from non-human primates naturally infected with *Trypanosoma cruzi* resemble the major immunological findings observed in human chagas disease. PLoS Neglected Tropical Diseases. 2016;**10**(1):e0004302. DOI: 10.1371/journal.pntd.0004302

[22] Vitelli-Avelar DM, Sathler-Avelar R, Mattoso-Barbosa AM, Gouin N, Perdigão-de-Oliveira M, Valério-dos-Reis L, et al. Cynomolgus macaques naturally infected with *Trypanosoma cruzi*-I exhibit an overall mixed pro-inflammatory modulated cytokine signature characteristic of human Chagas disease. PLoS Negl Trop Dis. 2017; 22;**11**(2):e0005233. DOI: 10.1371/journal.pntd.0005233

[23] Ferreira LRP, Frade AF, Baron MA, Navarro IC, Kalil J, Chevillard C, et al. Interferon-γ and other inflammatory mediators in cardiomyocyte signaling during Chagas disease cardiomyopathy. World Journal of Cardiology. 2014;**6**(8):782-790. DOI: 10.4330/wjc.v6.i8.782

[24] Sathler-Avelar R, Vitelli-Avelar DM, Massara RL, Borges JD, Lana M de, Teixeira-Carvalho A, et al. Benznidazole treatment during early-indeterminate Chagas' disease shifted the cytokine expression by innate and adaptive immunity cells toward a type 1-modulated immune profile. Scandinavian Journal of Immunology. 2006;**64**(5):554-563. DOI: 10.1111/j.1365-3083.2006.01843.x

[25] Orlic D, Kajstura J, Chimenti S, Jakoniuk I, Anderson SM, Li B, et al. Bone marrow cells regenerate infarcted myocardium. Nature. 2001;**410**(6829):701-705. DOI: 10.1038/35070587

[26] Sullivan NL, Eickhoff CS, Zhang X, Giddings OK, Lane TE, Hoft DF. Importance of the CCR5-CCL5 axis for mucosal *Trypanosoma cruzi* protection and B cell activation. Journal of Immunology. 2011;**187**(3):1358-1368. DOI: 10.4049/jimmunol.1100033

[27] Cardillo F, Postol E, Nihei J, Aroeira LS, Nomizo A, Mengel J. B cells modulate T cells so as to favour T helper type 1 and CD8+ T-cell responses in the acute phase of *Trypanosoma cruzi* infection. Immunology. 2007;**122**:584-595. DOI: 10.1111/j.1365-2567.2007.02677.x

[28] Zhang L, Tarleton RL. Characterization of cytokine production in murine *Trypanosoma cruzi* infection by in situ immunocytochemistry: Lack of association between susceptibility and Type 2 cytokine production. European Journal of Immunology. 1996;**26**:102-109

[29] Antunez MI, Cardoni RL. IL-12 and IFN-gamma production, and NK cell activity, in acute and chronic experimental *Trypanosoma cruzi* infections. Immunology Letters. 2000;**71**:103-109

[30] Vitelli-Avelar DM, Sathler-Avelar R, Teixeira-Carvalho A, Dias JCP, Gontijo ED, Faria AM, et al. Strategy to assess the overall cytokine profile of circulating leukocytes and its association with distinct clinical forms of human chagas disease. Scandinavian Journal of Immunology. 2008;**68**:516-525. DOI: 10.1111/j.1365-3083.2008.02167.x

[31] Magalhães LMD, Viana A, Chiari E, Galvão LMC, Gollob KJ, Dutra WO. Differential activation of human monocytes and lymphocytes by distinct strains of *Trypanosoma cruzi*. PLoS Neglected Tropical Diseases. 2015;**9**(7):e0003816. DOI: 10.1371/journal.pntd.0003816

[32] Roellig DM, Savage MY, Fujita AW, Barnabe C, Tibayrenc M, Steurer FJ, et al. Genetic variation and exchange in *Trypanosoma cruzi* isolates from the United States. PLoS One. 2013;**8**(2):e56198. DOI: 10.1371/journal.pone.0056198

[33] Zingales B, Miles MA, Campbell DA, Tibayrenc M, Macedo AM, Teixeira MM, et al. The revised *Trypanosoma cruzi* subspecific nomenclature: rationale, epidemiological relevance and research applications. Infection, Genetics and Evolution. 2012;**12**(2):240-253. DOI: 10.1016/j.meegid.2011.12.009

[34] Vago AR, Andrade LO, Leite AA, d'Avila Reis D, Macedo AM, Adad SJ, et al. Genetic characterization of *Trypanosoma cruzi* directly from tissues of patients with chronic Chagas disease: Differential distribution of genetic types into diverse organs. The American Journal of Pathology. 2000;**156**(5):1805-1809

[35] Freitas JM, Lages-Silva E, Crema E, Pena SD, Macedo AM. Real time PCR strategy for the identification of major lineages of *Trypanosoma cruzi* directly in chronically infected human tissues. International Journal for Parasitology. 2005;**35**(4):411-417. DOI: 10.1016/j.ijpara.2004.10.023

[36] Steindel M, Kramer Pacheco L, Scholl D, Soares M, de Moraes MH, Eger I, et al. Characterization of *Trypanosoma cruzi* isolated from humans, vectors, and animal reservoirs following an outbreak of acute human Chagas disease in Santa Catarina State, Brazil. Diagnostic Microbiology and Infectious Disease. 2008;**60**(1):25-32. DOI: 10.1016/j.diagmicrobio.2007.07.016

[37] Pissetti CW, Correia D, de Oliveira RF, Llaguno MM, Balarin MAS, Silva-Grecco RL, et al. Genetic and functional role of TNF-alpha in the development *Trypanosoma cruzi* infection. PLoS Neglected Tropical Diseases 2011;**5**(3):e976. DOI: 10.1371/journal.pntd.0000976

[38] Marcili A, Valente VC, Valente SA, Junqueira ACV, da Silva FM, Pinto AY das N, et al. *Trypanosoma cruzi* in Brazilian Amazonia: Lineages TCI and TCIIa in wild primates, *Rhodnius* spp. and in humans with chagas disease associated with oral transmission. International Journal for Parasitology. 2009;**39**(5):615-623. DOI: 10.1016/j.ijpara.2008.09.015

[39] Araújo CAC, Waniek PJ, Xavier SCC, Jansen AM. Genotype variation of *Trypanosoma cruzi* isolates from different Brazilian biomes. Experimental Parasitology. 2011;**127**(1):308-312. DOI: 10.1016/j.exppara.2010.07.013

[40] Lisboa CV, Mangia RH, Luz SLB, Kluczkovski A, Ferreira LF, Ribeiro CT, et al. Stable infection of primates with *Trypanosoma cruzi* I and II. Parasitology. 2006;**133**:603-611. DOI: 10.1017/S0031182006000722

[41] Bahia-Oliveira LMG, Gomes JAS, Cançado JR, Ferrari TC, Lemos EM, Luz ZMP, et al. Immunological and clinical evaluation of chagasic patients subjected to chemotherapy during the acute phase of *Trypanosoma cruzi* infection 14±30 years ago. The Journal of Infectious Diseases. 2000;**182**(2):634-638. DOI: 10.1086/315743

Chapter 9

The Endocannabinoid System in the Vervet Monkey Retina

Joseph Bouskila, Roberta Palmour,
Jean-François Bouchard and Maurice Ptito

Additional information is available at the end of the chapter

http://dx.doi.org/10.5772/intechopen.71830

Abstract

The main active compound found in the marijuana plant, tetrahydrocannabinol, is responsible for its psychotropic effects but also for its numerous beneficial actions such as appetite stimulation, nausea reduction, analgesia, and muscle spasm suppressor. Although cannabis consumption leads to some visual disturbances, the exact role of the endocannabinoid system (ECS) in normal vision is still unknown. Many studies have looked into the localization of this complex system (receptors, ligands, and enzymes) throughout the various components of the visual system of different animal models in order to obtain clues about its role. In fact, the retina, optic nerve, dorsal lateral geniculate nucleus, and visual cortices all express parts of the ECS. Manipulating this system pharmacologically or genetically has also an impact on visual function. In this book chapter, we provide the current understanding of how the ECS is involved in the functioning of the visual system and special emphasis is put on data obtained in monkeys, representing the most relevant animal model for visual neuroscience research. The mechanisms that control endocannabinoid (eCB) release and activation of cannabinoid receptors are discussed. We also propose a model highlighting the mechanisms involved in the regulation of photopic and scotopic vision taking advantage of the spatial specificity of the eCB signaling system and its physiological activation conditions.

Keywords: cannabinoids, retina, CB1 receptors, CB2 receptors, GPR55, vision, monkeys

1. Introduction

The medicinal use of cannabis can be traced back many thousands of years, but research on cannabinoids and the endocannabinoid system (ECS) was stimulated only in the mid-1960s after the isolation of tetrahydrocannabinol (THC) from the marijuana plant *Cannabis sativa* [1]. Later on, in the 1970s, several research groups reported independently that this compound is mostly responsible for the therapeutic and psychotropic effects of cannabis. Until the late 1980s, it was thought that cannabinoids would act nonspecifically on membranes and channels, but the discovery of specific receptors that bind to cannabinoids (THC and cannabidiol specifically) changed the course of events. A first cannabinoid receptor, termed cannabinoid receptor type 1 (CB1R), was discovered and cloned [2, 3]. Since the endogenous receptors do not stand and wait for cannabis consumption in order to get activated, the existence of endogenous molecules that activate the cannabinoid receptors was suggested. Indeed, the discovery of endogenous molecules that activate these receptors was made shortly after. Anandamide (AEA) and 2-arachidonoylglycerol (2-AG), the two most studied eCBs, then led to the discovery of specific enzymes that regulate the eCB levels. It is now clear that the ECS is present in many places in the organism and plays a neuromodulatory role at the cellular level. The eCB AEA and 2-AG bind to CB1R and CB2R with different affinities. There is also strong evidence that suggests that eCBs can target other receptors, particularly the putative "CB3" receptor GPR55 and the transient receptor potential vanilloid 1 (TRPV1) ion channel. Other eCB targets such as peroxisome proliferator-activated receptor (PPAR) and also CB1R are localized in the nucleus, where they shuttle from/to the cytosol in a ligand-dependent manner.

In addition to these receptors, the ECS is composed of various metabolic enzymes. The eCBs are produced "on demand" from membrane lipid precursors by multiple biosynthetic pathways. These bioactive lipid molecules are synthetized when and where needed and have important roles in physiological and pathophysiological conditions. AEA and 2-AG metabolism occurs through distinct routes that can overlap, of which several have been described in detail. The well-known view is that AEA is synthesized from membrane phospholipid precursors mainly by the action of N-acyl-phosphatidylethanolamine-specific phospholipase D (NAPE-PLD). By contrast, 2-AG is mainly synthesized by 2 diacylglycerol lipase enzymes, DAGLα and DAGLβ. The eCB-mediated effects are terminated by their fast degradation, mainly through the hydrolysis of AEA by the fatty acid amide hydrolase (FAAH) and of 2-AG by the monoacylglycerol lipase (MAGL). Besides these hydrolytic routes, AEA and 2-AG can also be oxidized by cyclooxygenase-2, distinct lipoxygenases, or cytochrome P450, all present in most tissues of the body. Interestingly, AEA and 2-AG oxidative by-products can also produce biological activity that may be mediated by different receptors.

2. Cannabis, the ECS, and the visual system

Besides the scleral vasodilation effect (also known as "red eye") of marijuana and the reduction of intraocular pressure [4, 5], the functional effects of cannabinoids on vision are still not well identified. A case study interviewing eight recent abstinent high-potency heavy cannabis smokers (approximately 56 g per month according to Ref. [6]) reported several categories of visual disturbances [7]. These included visual distortions, biased perception of distance, illusions of movement for stationary and moving objects, color intensification of objects, dimmed color, dimensional distortion, and blending of patterns and objects [7]. These visual illusions were also experienced by five patients with a history of previous use of marijuana [8]. Interestingly, pupil size, measured with a millimeter rule under constant illumination with eyes focused on an object at constant distance, is not changed after smoking marijuana [9, 10]. It has been shown that eCBs are present in ocular tissues, including the ciliary body, iris, choroid, as well as trabecular meshwork, but not on the lens [3, 5, 11–13]. Hence, eCBs may play an important role in eye function (such as regulation of intraocular pressure) under different normal and pathological conditions [14]. Furthermore, cannabis causes impaired performance in tests that require fine psychomotor control such as tracking a moving point of light on a screen [15, 16]. THC increases the time course of glare recovery by several seconds (5–10%) only at low contrast [16]. Higher doses of THC can produce side effects, including blur vision [17], double vision, and vision dimness [18]. Numerous reports claim that smoking marijuana improves dim light vision [19–21]. Acute consumption of marijuana reduces the Vernier and Snellen acuity, alters color discrimination, increases photosensitivity, and decreases dark adaptation [19, 21, 22]. No significant effect has been observed on static visual acuity [15] after consumption of THC with alcohol, although there was a marked reduction in acuity of moving targets when coordinated eye movements were required [23]. Binocular depth inversion is reduced in regular cannabis users while depth perception is not affected [23, 24]. Dronabinol, a synthetic THC, impaired binocular depth inversion and the top-down processing of visual sensory data [23]. Testing the visual functions by the use of steady-state visual evoked potential and electroencephalography over the occipital lobe suggests a disruption of later-stage visual processing in regular users [25].

3. Retinal anatomy and function in monkeys

3.1. The vervet monkey (*Chlorocebus sabaeus*)

The extensive resemblance of the nonhuman primates to *Homo sapiens* in various aspects, from genome sequence and molecular pathways to physiology and cognition, makes them the closest laboratory model to humans that cannot be approximated by any other animal model. Among the monkeys, old-world monkeys are the closest to the human physiology and behavior, after the apes. Old-world monkeys show more interspecies brain anatomy similarity compared with humans and apes [26]. Although the monkey brain is smaller, it has nonetheless a similar anatomical organization. For example, its visual system also comprises a retina, geniculostriate system, and the "what" (ventral) and "where" (dorsal) pathways, as in humans. The vervet monkey, or green African monkey, is an old-world monkey from the Cercopithecidae family native to Africa that 23 million years ago diverged from the hominoid family, and its genome is 96% homologous with man [27]. Our laboratories have been using vervets for many years now to particularly study the expression and function of the ECS in the retina and visual system. Like all the other old-world monkeys, vervets are medium to large size, have a tail with prehensile nerve ending, and are omnivorous with preference to plant matters. The St-Kitts vervet monkeys were imported from Senegambia in the seventeenth century [28–30]. Vervet monkeys are progressively chosen for biomedical research with a second citation record among the nonhuman primates after rhesus macaque [26]. Regarding the visual system, old-world monkeys have a foveal binocular vision with laminated retina with a high cone density that decreases with eccentricity and trichromatic color vision. The organization of the retinal mosaic has an impact on visual functions, the center being largely involved in visual acuity, color-coding, and photopic sensitivity (cone vision), whereas the periphery is more concerned with scotopic functions (rod vision) [31, 32]. Vervets also have a six-layered dorsal lateral geniculate nucleus (dLGN) and a laminar organization of the visual cortex similar to that seen in other old-world monkeys and humans [33, 34].

3.2. Description of the monkey retina

The fovea is a small central pit present in the surface of the retina in many types of fish, reptiles, and birds. Among mammals, primates are the only species with a *fovea centralis* [35]. The structure of the fovea can be slightly different in some types of animals. In some animals, the inner cell layers of the fovea may only show a reduced thickness, and in other animals, the fovea may have a complete absence of the same inner cell layers. In monkeys, cone photoreceptors line the base of the foveal pit and the other cells are displaced away from the foveal region. The fovea is at the intersection of retinal and optical axis of the globe and is the area of the most acute vision. The thickness of the retina is reduced in the fovea because the photoreceptor cell synapses and the inner retinal neurons are displaced peripherally from the foveal center. Cones are concentrated in the fovea; its center is free of rods. Peripherally, the number of rods increases, reaching a maximum at the perimeter of the fovea [36]. The diameter of the fovea in humans is 1.5 mm, and the central part, which is free of the retina inner layers, is 0.35 mm across and is called *foveola*. The tissue between the *foveola* and foveal rim, wedge-like in cross section, is called the *clivus* or foveal slope. The axons or outer fibers of the foveal rod and cone cells are elongated and form an additional layer, the "fiber layer of Henle," between the outer nuclear and outer plexiform layer in the periphery of the foveal area (**Figure 1**). The foveal photoreceptor cells, including

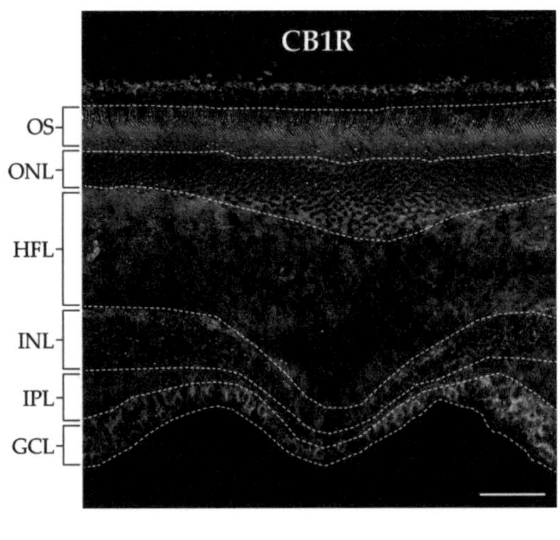

Figure 1. CB1R expression in the monkey fovea centralis. Confocal micrograph of a cross-section retina in the foveal pit immunolabeled for CB1R. OS, photoreceptor outer segments; ONL, outer nuclear layer; HFL, Henle fiber layer; INL, inner nuclear layer; IPL, inner plexiform layer; GCL, ganglion cell layer. Scale bar = 75 μm.

the Henle fibers, contain the yellow pigments zeaxanthin and lutein [37]. The area surrounding the fovea has a distinct tint. It is called *macula lutea*, the yellow spot. The words "macula" and "fovea" are frequently used as synonyms. Blood is supplied to the neuroretina, excluding photoreceptor cells, via retinal vessels. The large arteries and veins lie within the nerve fiber layer. Ascending arteries penetrate into the retinal tissue. Two flat beds of capillaries spread between the perikarya of horizontal cells and amacrine cells, at the outer and inner margins of the internal nuclear layer. Another network of capillaries supplies the ganglion cell and the nerve fiber layer.

3.3. The retinal endocannabinoid system in monkeys

The retina of many species expresses the ECS, including the tiger salamander, goldfish, mouse, rat, chick, tree shrew, vervet monkey, macaque monkey, and human. The anatomical organization of the retina in these species is obviously different, which makes it difficult to infer what really takes place in humans. For example, mice have a rod-dominated retina specialized for vision in nearly complete darkness, referred as scotopic conditions [38]. Additionally, tree shrews have a cone-dominated retina specialized for vision under well-lit conditions, referred to photopic conditions [39]. Primates, including monkeys, have a duplex retina, a fovea with a high cone density that decreases with eccentricity. Mouse and tree shrew retinas have no fovea compared to primates [40], and, compared to rodents, the retina of tree shrews is similar to primates [41, 42]. Comparative studies on the organization of the retina of different animal species led to the conclusion that ancestral mammals may have already developed cone photopigments [32]. Many components of the ECS have been localized in cone photoreceptors, horizontal, amacrine, bipolar, and retinal ganglion cells in the central and peripheral retina of vervet monkeys (**Figure 1**; [43]).

Compared to rodents, the retina of primates including monkeys and humans has the unique characteristic to have a duplex retina with a cone-dominated fovea [44]. As part of the brain, this highly organized tissue processes visual information in parallel channels. While the input retina consists of only 2 types of photoreceptors (rods sensitive to 1 wavelength of light and cones selective to 3 different wavelengths), the output retina contains more than 20 types of ganglion cells [45, 46]. The primate retina exhibits a strikingly high expression of CB1R, the main cannabinoid-binding protein responsible for the marijuana psychotropic effects. Cones of the central retina abundantly expressed CB1R. The vertical glutamate pathway (cone photoreceptors-bipolar cells-ganglion cells) also heavily expresses CB1R. While the functional importance of retinal CB1R is supported by anatomical data in rodents and primates, evidence for a role of eCBs in synaptic signaling is provided by *in vivo* ERG experiments on vervet monkeys. The presence of CB1 receptors in the retina of many species has been reported [47]. The modulatory effects of cannabinoids, acting on CB1 receptors, at all stages of retinal processing have also been described (For review, see [48]). Furthermore, the cannabinoid receptors (**Figure 2**) and related enzymes (**Figure 3**) are expressed in the mouse, tree shrew, vervet monkey, and macaque monkey retina [49]. More specifically, the ECS is present throughout the monkey retina, from the foveal pit to the periphery, suggesting that it may play a role in retinal function.

3.4. Electroretinography in monkeys

Electroretinography (ERG) is a noninvasive ocular test that measures the electrical responses of various cell types in the retina, including rod and cone photoreceptors, horizontal cells, bipolar cells, amacrines cells, ganglion cells, and Müller cells. This diagnostic tool can objectively evaluate retinal function in clinical and research settings. It is well-known that the monkey eye anatomy and physiology are similar to those of humans, making it the preferred nonhuman primate animal model for testing ocular effects. Given that the technical and procedural aspects of the human

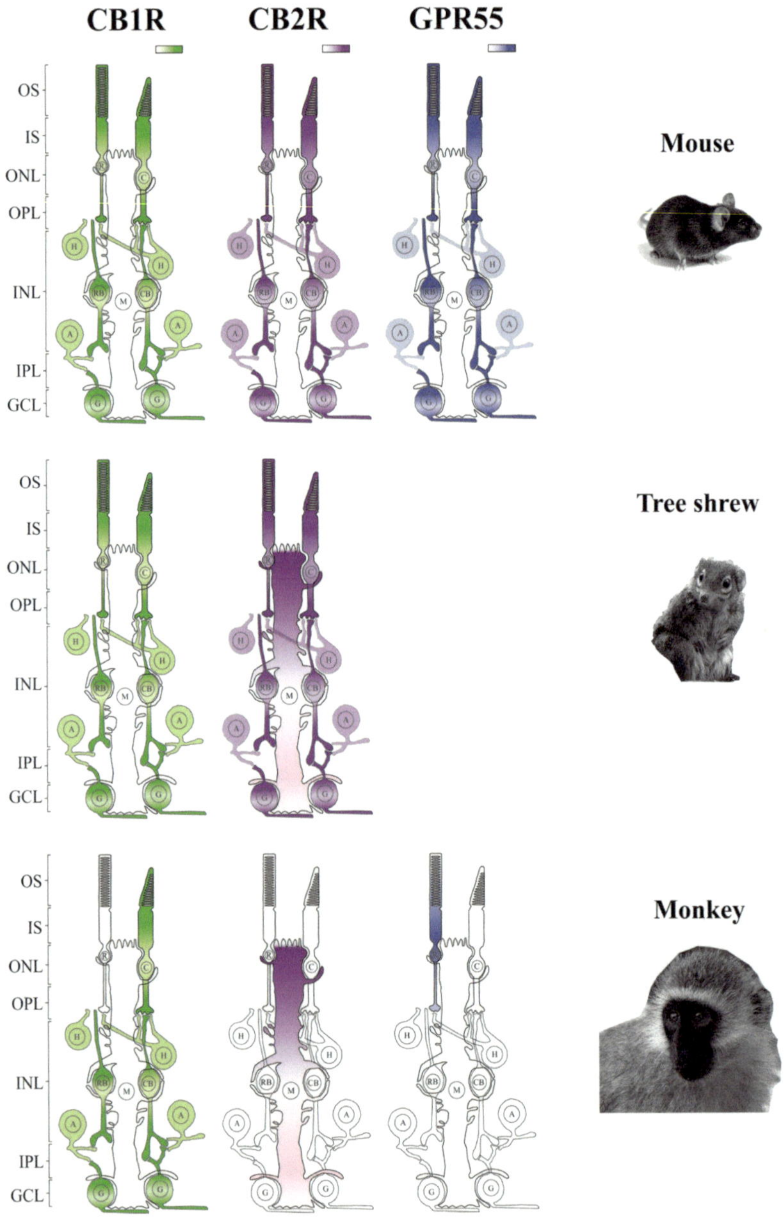

Figure 2. Mapping of the receptors CB1R, CB2R, and GPR55 in the mouse, tree shrew, and monkey retina. These receptors are differently expressed in the retina of these mammals. These results are compiled from several published articles [43, 47–54]. OS, photoreceptor outer segments; IS, photoreceptor inner segments; ONL, outer nuclear layer; ONL, outer plexiform layer; INL, inner nuclear layer; IPL, inner plexiform layer; GCL, ganglion cell layer.

protocol have been standardized by the International Society for Clinical Electrophysiology of Vision [58], the ERG can be routinely used to assess toxicity (potential global neurotoxicity induced by drugs in primates) on retinal function. With this standard method, a tremendous amount of

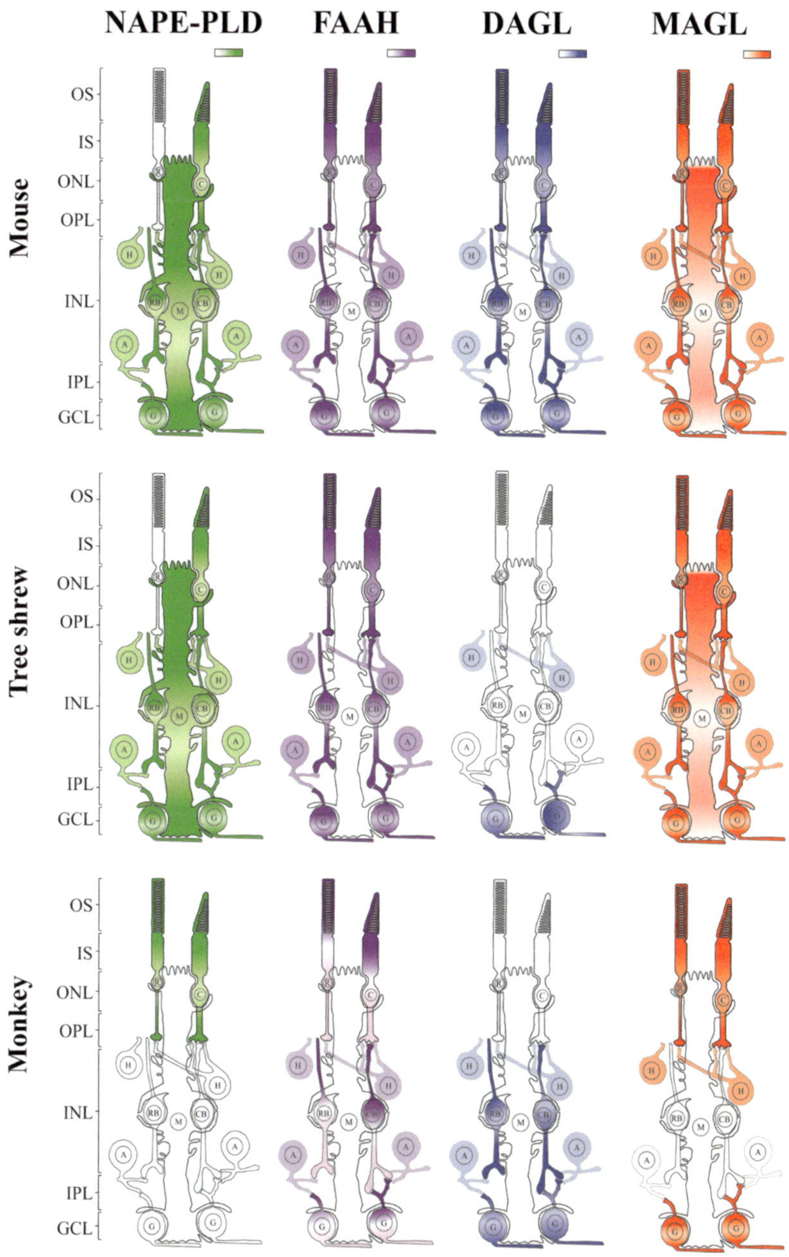

Figure 3. Localization of the cannabinoid-related enzymes in the mouse, tree shrew, and monkey retina. NAPE-PLD, FAAH, DAGL, and MAGL expressions have some similarities and differences in the retina of these mammals. These results summarize results published in several articles [43, 49, 50, 55–57]. OS, photoreceptor outer segments; IS, photoreceptor inner segments; ONL, outer nuclear layer; ONL, outer plexiform layer; INL, inner nuclear layer; IPL, inner plexiform layer; GCL, ganglion cell layer.

data can be accumulated for eye physiopathology evaluation. This noninvasive, nonpainful, nondamaging technique does not induce undesirable effects to the eye. Numerous studies of ocular toxicity in macaque monkeys have produced a standard ERG protocol for the assessment of retinal function [59]. This method allows independent testing of rod and cone systems. If a low-intensity flash stimulus is conveyed to the dark-adapted retina, the rod system is stimulated. If a flash stimulus is conveyed to the light-adapted retina, the cone system is targeted. When strong flash stimuli are elicited, the retina electrical responses will produce an ERG waveform comprising an initial corneal-negative deflection derived from rods and cones, the a-wave, followed by a corneal-positive deflection derived from the inner retina (predominantly Müller and ON-bipolar cells), the b-wave. A standardized experimental protocol was developed in our laboratory and used to study the role of the ECS on retinal function in vervet monkeys [60]. The mobile experimental setup consisted of placing ERG-Jet electrodes in both eyes for simultaneous recordings (**Figure 4**). In dark-adapted conditions, dim flashes activate the scotopic system (rod pathway) and photopic flashes evoke a mixed response (rod and cone pathways). In light-adapted conditions, flashes produce only a photopic system response (cone pathway).

CB1R, CB2R, and GPR55 play an important role in retinal function. The function of the cannabinoid receptors in the retina has been highlighted in ERG studies in adult mice [53] and vervet monkeys [61, 62]. Cécyre et al. [53] demonstrated a significant change in the ERG a-wave of the CB2R knockout dark-adapted mouse but not in CB1R knockouts compared to wild types. They concluded that CB2R is likely to play a greater role in mouse retinal processing than CB1R. In monkeys, we have recently reported a significant increase in the b-wave component of the scotopic and photopic ERG after a blockade of both CB1R and CB2R with their specific antagonists [61]. This variation can be due to the different pattern of expression and specialization of the ECS in the monkey. We also reported that GPR55 may play an instrumental role in mediating scotopic vision, because it is exclusively expressed in rods [52] and it modulates specifically scotopic retinal function [62]. **Figure 5** summarizes the ERG effects

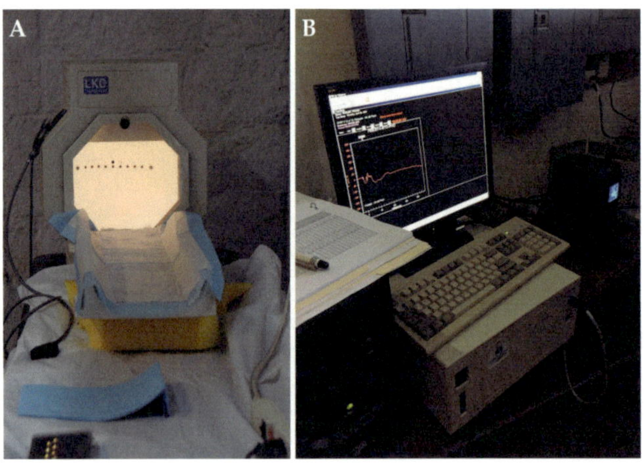

Figure 4. Experimental setup for ERG recordings in vervet monkeys. (A) The ganzfeld allows us to illicit full-field flashes. (B) The ERG machine (UTAS-E3000) is linked to a computer.

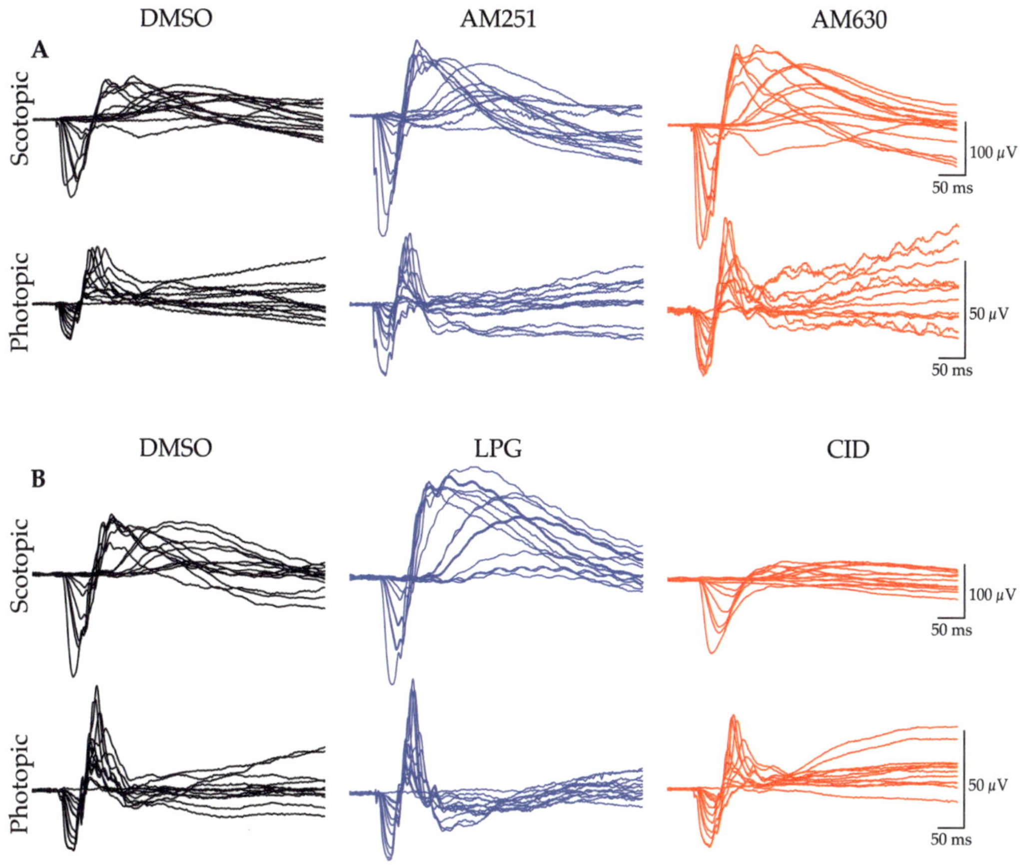

Figure 5. The effect of modulating CB1R, CB2R, or GPR55 in the monkey retina. (A) The intravitreal injection of AM251, an inverse agonist of CB1R, or AM630, an inverse agonist of CB2R, causes an increase of the scotopic and photopic responses compared to the vehicle, dimethyl sulfoxide (DMSO). (B) The intravitreal injection of lysophosphatidylglucoside (LPG), an agonist of GPR55, causes an increase of the scotopic response, but not of the photopic response. Conversely, the intravitreal injection of CID16020046 (CID), an antagonist of GPR55, causes a decrease of the scotopic response, but not of the photopic response.

obtained in vervet monkeys. This body of evidence (the anecdotal reports, the anatomical localization of the ECS, and its functional implications) indicates that eCBs are involved in shaping retinal responses to light and suggests it plays a crucial role in visual processing.

3.5. The eCB signaling pathways in the monkey retina

The presence of CB1R in the neuroretina (in the vertical pathway consisting of photoreceptors, bipolar cells, and ganglion cells), of CB2R in the major glial element of the retina (Müller cells), and of GPR55 in rod photoreceptors suggest differential retinal function. Furthermore, localization of the metabolic enzymes suggests that eCBs are synthesized and released in the synapse surrounding the neurons from which they are released. They therefore act locally on adjacent retinal cells [63]. This could in turn influence, directly (through CB1R or GPR55) or indirectly (through CB2R), the release of glutamate, the main neurotransmitter of the retinal vertical pathway. After the ligands are produced, many ionic channels such as K^+ and Ca^{++} are modulated after activation of the cannabinoid receptors.

The suggested hypothetical function of the ECS in the monkey retina may be as follows. In photopic conditions, when cones are stimulated by light, the ionic channels are inhibited, a process known as the *"inhibition of the retinal dark currents."* The resulting phototransduction reduces the glutamate release in the synapse and propagates an evoked potential to bipolar cells. Given the localization of the metabolic enzymes in monkeys (**Figure 3**), the same bipolar cells may be the main source of eCB production that will act in a retrograde manner and activate CB1R located in cone pedicles, thus regulating glutamate release. This eCB production will also synthesize 2-AG that will activate CB2R in Müller cells, thus modulating potassium spatial buffering throughout the retina. The activation of CB2R coupled to $G_{i/o}$ will reduce the levels of cyclic AMP and PKA ([64] for review; [65]). Given that PKA activates $K_{IR}4.1$ channels in Müller cells [66], CB2R will play a role by negatively modulating potassium.

In scotopic conditions, the synaptic terminals of rods release a large quantity of glutamate. This glutamate binds to mGluR6 receptors located on the dendrites of ON rod bipolar cells [67]. Activation of GPR55 by its endogenous agonist (lysophosphatidylglucoside [LPG]) will stimulate the $G_{\alpha13}$, RhoA, ROCK, and PLC cascade to open Na^+/Ca^{++} channels, induce a membrane depolarization, and, finally, modify the scotopic ERG [52, 62].

These proposed mechanisms of action for the photoreceptor-bipolar cell synapse are illustrated in **Figure 6** and could also take place in other synapses in the primate retina. However, even though great efforts are deployed to understand the precise function of the retinal ECS, behavioral studies are crucial to clearly establishing its role in vision.

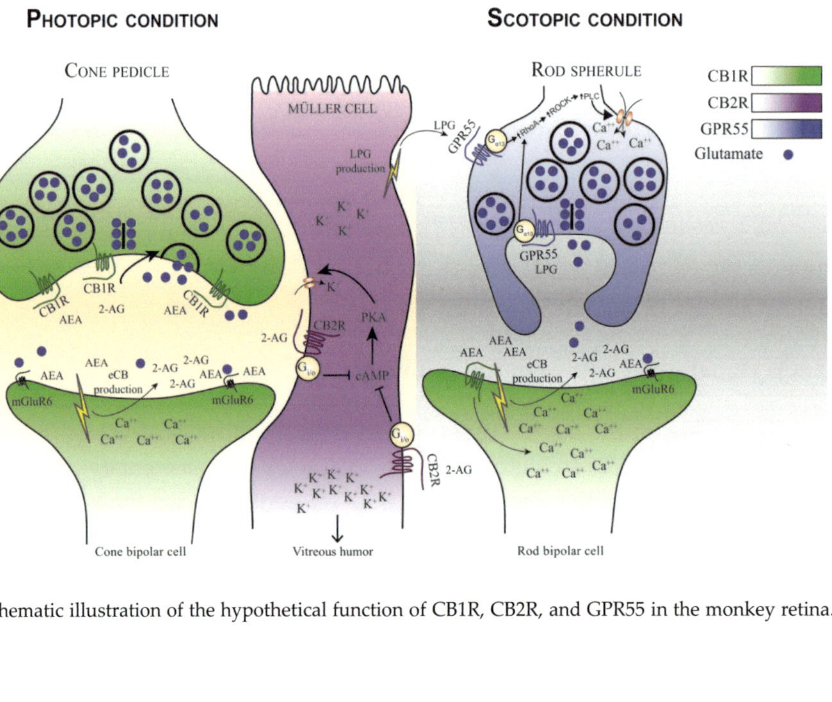

Figure 6. Schematic illustration of the hypothetical function of CB1R, CB2R, and GPR55 in the monkey retina. See text for details.

http://dx.doi.org/10.5772/intechopen.71830

4. Beyond the retina, in the visual thalamus and visual cortex

Multiple studies have studied the expression of the ECS in the thalamus of rodents and primates, but only one has thoroughly examined the expression of CB1R, NAPE-PLD, and FAAH in the dorsal lateral geniculate nucleus of vervet monkeys [34]. One study has characterized the physiological effects of CB1R-mediated activity in the visual thalamus of rats [68]. Using single-unit extracellular recordings at the level of the thalamus, CB1R activation affected two cell populations: one exhibiting excitatory effects (28%) and the other inhibitory effects (72%) [68]. When AM251 (inverse agonist of CB1R) was added, these actions were blocked. The authors concluded that CB1R in the rat thalamus acts as a dynamic modulator of visual information. Furthermore, the CB1R system is present in the lateral geniculate nucleus of monkeys [34]. The presence of CB1R in the dLGN of vervet monkeys could have a higher impact on visual processing, compared to what is found in rats [68], due to its complex laminar structure and increased retinal inputs. The results reported by Javadi et al. [34] are in accordance with these electrophysiological findings and revealed that this neuromodulation of CB1R receptors may be due to their expression in the dLGN. Even though the expression of CB1R and FAAH is more abundant in magnocellular layers, they are nonetheless present in the parvocellular layers of the dLGN. The high expression of CB1R in the magnocellular layers may explain some of the behavioral effects of cannabis and cannabinoids associated with the integrity of the dorsal visual pathway (visual-spatial localization and motion perception) [34]. If the ECS plays any role in color perception, it should be through chromatic properties of parvo cells and retinal cones. Both of these visual system components express CB1R. Conjointly, one of the most frequently reported cannabis effects is more intense and brighter colors [7, 69].

The visual cortex is important to the conscious processing of visual information. The exact localization of the ECS in higher visual structures remains somewhat unclear. Several published reports describe the effects of cannabinoids on visual perception (thresholds of light detection, glare recovery, and color vision). The localization of the ECS within the higher order visual areas responsible for these changes in perception is starting to be revealed. As noted above, there has been evidence for central effects of cannabinoid use in vision by binocular depth inversion technique and EEG recordings of the occipital cortex [24, 25]. Yoneda and colleagues reported that the expression and localization of CB1R in the visual cortex of the mouse is regulated during the development and through visual experiences. In mouse, expression of CB1R in deep layers of V1 decreased after dark rearing from birth to P30. However, 2 days of monocular deprivation upregulated the localization of CB1R in inhibitory nerve terminal in deep layers [70]. It has also been reported that CP55940, a full agonist of CB1R and CB2R, decreases the electroencephalogram power, and the local field power and coherence, in V1 and V2 in macaque monkeys [71].

5. The ECS and visual disorders

The ECS modulates many neurotransmission processes in the central nervous system. In fact, numerous recent researches raised the impact of metabotropic and ionotropic receptors on neurological diseases. Studying modulator systems like the monoaminergic, purinergic, and cholinergic systems may reveal the pathophysiology of many disorders. Opioid and nicotinic receptors were generally analyzed in order to treat drugs of abuse, morphine and nicotine. Conversely, the study of cannabinoid receptors is not primarily to treat addiction to marijuana but has a much broader role. Indeed, CB1R is the most abundant metabotropic receptor in the CNS [72]. These receptors together with their ligands and related enzymes constitute a goldmine in the chase of finding therapeutic targets against many visual pathologies. For instance, blindness and visual impairment are relatively refractory to most of the current drugs, emphasizing the importance of identifying a novel site of action for pharmacological treatments. Accordingly, modulation of the cannabinoid system remains potentially a new therapeutic approach. This could be performed at several levels. For endogenous cannabinoids, it would be a modulation of their synthesis, release, cellular uptake, metabolism, or interactions with cannabinoid receptors. Biochemical imbalances in the ECS in visual structures may cause or exacerbate pathological disorders, such as glaucoma ([73] for review), diabetic retinopathy, and age-related macular degeneration [14]. The variation of the content of eCBs in these diseases suggests that they play a crucial role in ocular homeostasis. Indeed, patients with glaucoma have a decrease in 2-AG levels in the ciliary body [12]. In age-related macular degeneration (AMD) patients, 2-AG levels are amplified in the iris and AEA is also increased in the retina [14]. The same pattern of augmentation of AEA was observed in choroid, ciliary body, and cornea of the AMD patients [14]. Moreover, many recent studies investigating the role of cannabinoids in visual development have shown that CB1R is transiently expressed throughout development of the chick and rat retinas. Nevertheless, cannabis and cannabinoids as therapeutic agents have not yet been unequivocally established. Targeting preferentially retinal cannabinoid receptors to avoid unwanted psychotropic effects is a new interesting avenue requiring further investigation.

6. Future research directions

There is now strong evidence that suggests that the ECS plays a significant role in regulating visual function. However, several important questions remain to be answered. First, there is a need to define the exact mechanisms by which eCB production in the retina is regulated. Then, we need to determine if eCB production is influenced by classical calcium-regulating hormones, cytokines, and mechanical loading. Further research is also necessary to define the exact signaling pathways used by cannabinoid receptors to regulate retinal cell activity. There is evidence that CB1R regulates photoreceptors, horizontal cells, bipolar cells, amacrine cells, and ganglion cell activity through a cAMP-mediated pathway (for review, see [48]), but little is known on how cannabinoids regulate global retinal activity. There have been major inconsistencies between different studies with regard to the expression and function of the ECS in the retina and the brain. These inconsistencies may be due to the nonspecificity of the available cannabinoid receptor ligands that are thought to be specific for CB1R or CB2R but can actually bind to GPR55 [74–76]. A further area of research that remains to be explored is to determine how cannabinoid receptors exert their effects on vision, through a neuronal or glial mechanism. This is clinically relevant since if the glial effects were predominant, it may be possible to develop agonists of these receptors that do not affect neuronal function but could favorably influence visual function without causing adverse psychotropic neuronal effects. The outcome of these studies will greatly enhance our understanding of the role of the ECS in vision and encourage the development of new treatments for visual disorders based on targeting the ECS.

Acknowledgements

Our studies reported in this book chapter have all been supported by grants from the Natural Sciences and Engineering Research Council of Canada to Joseph Bouskila (postdoctoral fellowship), Jean-François Bouchard, and Maurice Ptito and from the Canadian Institutes of Health Research to Roberta Palmour, Maurice Ptito and Jean-François Bouchard.

Author details

Joseph Bouskila[1,2], Roberta Palmour[1], Jean-François Bouchard[2] and Maurice Ptito[2,3]*

*Address all correspondence to: maurice.ptito@umontreal.ca

1 Department of Human Genetics, McGill University, Montreal, Quebec, Canada

2 School of Optometry, University of Montreal, Montreal, Quebec, Canada

3 Laboratory of Neuropsychiatry-Psychiatric Centre, University of Copenhagen, Copenhagen, Denmark

References

[1] Gaoni Y, Mechoulam R. Isolation, structure, and partial synthesis of an active constituent of hashish. Journal of the American Chemical Society. 1964;**86**(8):1646

[2] Devane WA, Hanus L, Breuer A, Pertwee RG, Stevenson LA, Griffin G, et al. Isolation and structure of a brain constituent that binds to the cannabinoid receptor. Science. 1992;**258**(5090):1946-1949

[3] Matsuda LA, Lolait SJ, Brownstein MJ, Young AC, Bonner TI. Structure of a cannabinoid receptor and functional expression of the cloned cDNA. Nature. 1990;**346**(6284):561-564

[4] Green K. The ocular effects of cannabinoids. Current Topics in Eye Research. 1979;**1**: 175-215

[5] Porcella A, Casellas P, Gessa GL, Pani L. Cannabinoid receptor CB1 mRNA is highly expressed in the rat ciliary body: Implications for the antiglaucoma properties of marihuana. Molecular Brain Research. 1998;**58**(1-2):240-245

[6] Schwartz RH, Hayden GF, Riddile M. Laboratory detection of marijuana use: Experience with a photometric immunoassay to measure urinary cannabinoids. American Journal of Diseases of Children. 1985;**139**(11):1093-1096

[7] Lerner AG, Goodman C, Rudinski D, Bleich A. Benign and time-limited visual disturbances (flashbacks) in recent abstinent high-potency heavy cannabis smokers: A case series study. The Israel Journal of Psychiatry and Related Sciences. 2011;**48**(1):25-29

[8] Levi L, Miller NR. Visual illusions associated with previous drug abuse. Journal of Clinical Neuro-Ophthalmology. 1990;**10**(2):103-110

[9] Weil AT, Zinberg NE, Nelsen JM. Clinical and psychological effects of marijuana in man. The International Journal of the Addictions. 1969;**4**(3):427-451

[10] Brown B, Adams AJ, Haegerstrom-Portnoy G, Jones RT, Flom MC. Pupil size after use of marijuana and alcohol. American Journal of Ophthalmology. 1977;**83**(3):350-354

[11] Porcella A, Maxia C, Gessa GL, Pani L. The human eye expresses high levels of CB1 cannabinoid receptor mRNA and protein. The European Journal of Neuroscience. 2000;**12**(3):1123-1127

[12] Chen J, Matias I, Dinh T, Lu T, Venezia S, Nieves A, et al. Finding of endocannabinoids in human eye tissues: Implications for glaucoma. Biochemical and Biophysical Research Communications. 2005;**330**(4):1062-1067

[13] Stumpff F, Boxberger M, Krauss A, Rosenthal R, Meissner S, Choritz L, et al. Stimulation of cannabinoid (CB1) and prostanoid (EP2) receptors opens BKCa channels and relaxes ocular trabecular meshwork. Experimental Eye Research. 2005;**80**(5):697-708

[14] Matias I, Wang JW, Moriello AS, Nieves A, Woodward DF, Di Marzo V. Changes in endocannabinoid and palmitoylethanolamide levels in eye tissues of patients with diabetic retinopathy and age-related macular degeneration. Prostaglandins, Leukotrienes, and Essential Fatty Acids. 2006;**75**(6):413-418

[15] Adams AJ, Brown B, Flom MC, Jones RT, Jampolsky A. Alcohol and marijuana effects on static visual acuity. American Journal of Optometry and Physiological Optics. 1975;**52**(11):729-735

[16] Adams AJ, Brown B, Haegerstrom-Portnoy G, Flom MC, Jones RT. Marijuana, alcohol, and combined drug effects on the time course of glare recovery. Psychopharmacology. 1978;**56**(1):81-86

[17] Noyes R Jr, Brunk SF, Avery DA, Canter AC. The analgesic properties of delta-9-tetrahydrocannabinol and codeine. Clinical Pharmacology and Therapeutics. 1975;**18**(1):84-89

[18] Consroe P, Musty R, Rein J, Tillery W, Pertwee R. The perceived effects of smoked cannabis on patients with multiple sclerosis. European Neurology. 1997;**38**(1):44-48

[19] Dawson WW, Jimenez-Antillon CF, Perez JM, Zeskind JA. Marijuana and vision—After ten years' use in Costa Rica. Investigative Ophthalmology & Visual Science. 1977;**16**(8):689-699

[20] Merzouki A, Mesa JM. Concerning kif, a *Cannabis sativa* L. preparation smoked in the Rif mountains of northern Morocco. Journal of Ethnopharmacology. 2002;**81**(3):403-406

[21] Russo EB, Merzouki A, Mesa JM, Frey KA, Bach PJ. Cannabis improves night vision: A case study of dark adaptometry and scotopic sensitivity in kif smokers of the Rif mountains of northern Morocco. Journal of Ethnopharmacology. 2004;**93**(1):99-104

[22] Kiplinger GF, Manno JE, Rodda BE, Forney RB. Dose-response analysis of the effects of tetrahydrocannabinol in man. Clinical Pharmacology and Therapeutics. 1971;**12**(4): 650-657

[23] Leweke FM, Schneider U, Thies M, Munte TF, Emrich HM. Effects of synthetic delta9-tetrahydrocannabinol on binocular depth inversion of natural and artificial objects in man. Psychopharmacology. 1999;**142**(3):230-235

[24] Semple DM, Ramsden F, McIntosh AM. Reduced binocular depth inversion in regular cannabis users. Pharmacology, Biochemistry, and Behavior. 2003;**75**(4):789-793

[25] Skosnik PD, Krishnan GP, Vohs JL, O'Donnell BF. The effect of cannabis use and gender on the visual steady state evoked potential. Clinical Neurophysiology. 2006;**117**(1):144-156

[26] Jasinska AJ, Schmitt CA, Service SK, Cantor RM, Dewar K, Jentsch JD, et al. Systems biology of the vervet monkey. ILAR Journal. 2013;**54**(2):122-143

[27] Goodman M, Porter CA, Czelusniak J, Page SL, Schneider H, Shoshani J, et al. Toward a phylogenetic classification of primates based on DNA evidence complemented by fossil evidence. Molecular Phylogenetics and Evolution. 1998;**9**(3):585-598

[28] Labat JB. Nouveau Voyage Aux Isles de l'Amérique: P. Husson; 1722

[29] McGuire MT. The St. Kitts Vervet. Basel: Karger; 1974

[30] Palmour RM, Mulligan J, Howbert JJ, Ervin F. Of monkeys and men: Vervets and the genetics of human-like behaviors. American Journal of Human Genetics. 1997;**61**(3):481-488

[31] Wässle H, Grüunert U, Chun M-N, Boycott BB. The rod pathway of the macaque monkey retina: Identification of AII-amacrine cells with antibodies against calretinin. The Journal of Comparative Neurology. 1995;**361**(3):537-551

[32] Jacobs GH. Primate color vision: A comparative perspective. Visual Neuroscience. 2008;**25**(5-6):619-633

[33] Burke MW, Kupers R, Ptito M. Adaptive neuroplastic responses in early and late hemispherectomized monkeys. Neural Plasticity. 2012;**2012**:852423

[34] Javadi P, Bouskila J, Bouchard JF, Ptito M. The endocannabinoid system within the dorsal lateral geniculate nucleus of the vervet monkey. Neuroscience. 2015;**288**:135-144

[35] Krebs W, Krebs I. Primate Retina and Choroid: Atlas of Fine Structure in Man and Monkey. Springer-Verlag; 1991

[36] Osterberg G. Topography of the layer of rods and cones in the human retina. Acta Ophthalmologica. Supplementum. 1935;**6**:11-97

[37] Bone RA, Landrum JT, Fernandez L, Tarsis SL. Analysis of the macular pigment by HPLC: Retinal distribution and age study. Investigative Ophthalmology & Visual Science. 1988;**29**(6):843-849

[38] Jeon CJ, Strettoi E, Masland RH. The major cell populations of the mouse retina. The Journal of Neuroscience. 1998;**18**(21):8936-8946

[39] Müller B, Peichl L. Horizontal cells in the cone-dominated tree shrew retina: Morphology, photoreceptor contacts, and topographical distribution. The Journal of Neuroscience. 1993;**13**(8):3628-3646

[40] Prusky GT, Douglas RM. Characterization of mouse cortical spatial vision. Vision Research. 2004;**44**(28):3411-3418

[41] Müller B, Peichl L. Topography of cones and rods in the tree shrew retina. The Journal of Comparative Neurology. 1989;**282**(4):581-594

[42] Fan Y, Huang ZY, Cao CC, Chen CS, Chen YX, Fan DD, et al. Genome of the Chinese tree shrew. Nature Communications. 2013;**4**:1426

[43] Bouskila J, Burke MW, Zabouri N, Casanova C, Ptito M, Bouchard JF. Expression and localization of the cannabinoid receptor type 1 and the enzyme fatty acid amide hydrolase in the retina of vervet monkeys. Neuroscience. 2012;**202**:117-130

[44] Frishman LJ, Wang MH. Chapter 24—Electroretinogram of human, monkey and mouse. In: Levin LA, Nilsson SFE, Hoeve JV, Wu S, Kaufman PL, Alm A, editors. Adler's Physiology of the Eye. 11th ed. Elsevier Health Sciences; 2011. pp. 480-501

[45] Masland RH. The fundamental plan of the retina. Nature Neuroscience. 2001;**4**(9):877-886

[46] Masland RH. Neuronal diversity in the retina. Current Opinion in Neurobiology. 2001;**11**(4):431-436

[47] Straiker AJ, Maguire G, Mackie K, Lindsey J. Localization of cannabinoid CB1 receptors in the human anterior eye and retina. Investigative Ophthalmology & Visual Science. 1999;**40**(10):2442-2448

[48] Yazulla S. Endocannabinoids in the retina: From marijuana to neuroprotection. Progress in Retinal and Eye Research. 2008;**27**(5):501-526

[49] Bouskila J, Javadi P, Elkrief L, Casanova C, Bouchard JF, Ptito M. A comparative analysis of the Endocannabinoid system in the retina of mice, tree shrews, and monkeys. Neural Plasticity 2016;**2016**:3127658

[50] Hu SS, Arnold A, Hutchens JM, Radicke J, Cravatt BF, Wager-Miller J, et al. Architecture of cannabinoid signaling in mouse retina. The Journal of Comparative Neurology. 2010;**518**(18):3848-3866

[51] Bouskila J, Javadi P, Casanova C, Ptito M, Bouchard JF. Muller cells express the cannabinoid CB2 receptor in the vervet monkey retina. The Journal of Comparative Neurology. 2013;**521**(11):2399-2415

[52] Bouskila J, Javadi P, Casanova C, Ptito M, Bouchard JF. Rod photoreceptors express GPR55 in the adult vervet monkey retina. PLoS One. 2013;**8**(11):e81080

[53] Cécyre B, Zabouri N, Huppé-Gourgues F, Bouchard JF, Casanova C. Roles of cannabinoid receptors type 1 and 2 on the retinal function of adult mice. Investigative Ophthalmology & Visual Science. 2013;**54**(13):8079-8090

[54] Cherif H, Argaw A, Cecyre B, Bouchard A, Gagnon J, Javadi P, et al. Role of GPR55 during axon growth and target innervation. eNeuro. 2015;**2**(5). DOI: https://doi.org/10.1523/ENEURO.0011

[55] Cécyre B, Monette M, Beudjekian L, Casanova C, Bouchard JF. Localization of diacylglycerol lipase alpha and monoacylglycerol lipase during postnatal development of the rat retina. Frontiers in Neuroanatomy. 2014;**8**:150

[56] Zabouri N, Bouchard JF, Casanova C. Cannabinoid receptor type 1 expression during postnatal development of the rat retina. The Journal of Comparative Neurology. 2011;**519**(7):1258-1280

[57] Zabouri N, Ptito M, Casanova C, Bouchard JF. Fatty acid amide hydrolase expression during retinal postnatal development in rats. Neuroscience. 2011;**195**:145-165

[58] McCulloch DL, Marmor MF, Brigell MG, Hamilton R, Holder GE, Tzekov R, et al. ISCEV standard for full-field clinical electroretinography (2015 update). Documenta Ophthalmologica. 2015;**130**(1):1-12

[59] Bee WH. Standardized electroretinography in primates: A non-invasive preclinical tool for predicting ocular side effects in humans. Current Opinion in Drug Discovery & Development. 2001;**4**(1):81-91

[60] Bouskila J, Javadi P, Palmour RM, Bouchard JF, Ptito M. Standardized full-field electroretinography in the green monkey (Chlorocebus Sabaeus). PLoS One. 2014;**9**(10):e111569

[61] Bouskila J, Harrar V, Javadi P, Beierschmitt A, Palmour R, Casanova C, et al. Cannabinoid receptors CB1 and CB2 modulate the electroretinographic waves in Vervet monkeys. Neural Plasticity. 2016;**2016**:1253245

[62] Bouskila J, Harrar V, Javadi P, Casanova C, Hirabayashi Y, Matsuo I, et al. Scotopic vision in the monkey is modulated by the G protein-coupled receptor 55. Visual Neuroscience. 2016;**33**:E006

[63] Gómez-Ruiz M, Hernández M, de Miguel R, Ramos JA. An overview on the biochemistry of the cannabinoid system. Molecular Neurobiology. 2007;**36**(1):3-14

[64] Howlett AC, Barth F, Bonner TI, Cabral G, Casellas P, Devane WA, et al. International Union of Pharmacology. XXVII. Classification of cannabinoid receptors. Pharmacological Reviews. 2002;**54**(2):161-202

[65] Bolognini D, Costa B, Maione S, Comelli F, Marini P, Di Marzo V, et al. The plant cannabinoid Delta9-tetrahydrocannabivarin can decrease signs of inflammation and inflammatory pain in mice. British Journal of Pharmacology. 2010;**160**(3):677-687

[66] MacGregor GG, Xu JZ, McNicholas CM, Giebisch G, Hebert SC. Partially active channels produced by PKA site mutation of the cloned renal K+ channel, ROMK2 (kir1.2). The American Journal of Physiology. 1998;**275**(3 Pt 2):F415-F422

[67] Sampath AP, Rieke F. Selective transmission of single photon responses by saturation at the rod-to-rod bipolar synapse. Neuron. 2004;**41**(3):431-443

[68] Dasilva MA, Grieve KL, Cudeiro J, Rivadulla C. Endocannabinoid CB1 receptors modulate visual output from the thalamus. Psychopharmacology. 2012;**219**(3):835-845

[69] Green B, Kavanagh D, Young R. Being stoned: A review of self-reported cannabis effects. Drug and Alcohol Review. 2003;**22**(4):453-460

[70] Yoneda T, Kameyama K, Esumi K, Daimyo Y, Watanabe M, Hata Y. Developmental and visual input-dependent regulation of the CB1 cannabinoid receptor in the mouse visual cortex. PLoS One. 2013;**8**(1):e53082

[71] Ohiorhenuan IE, Mechler F, Purpura KP, Schmid AM, Hu Q, Victor JD. Cannabinoid neuromodulation in the adult early visual cortex. PLoS One. 2014;**9**(2):e87362

[72] Herkenham M, Lynn AB, Johnson MR, Melvin LS, de Costa BR, Rice KC. Characterization and localization of cannabinoid receptors in rat brain: A quantitative in vitro autoradiographic study. The Journal of Neuroscience. 1991;**11**(2):563-583

[73] Cairns EA, Toguri JT, Porter RF, Szczesniak AM, Kelly ME. Seeing over the horizon—Targeting the endocannabinoid system for the treatment of ocular disease. Journal of Basic and Clinical Physiology and Pharmacology. 2016;**27**(3):253-265

[74] Kapur A, Zhao P, Sharir H, Bai Y, Caron MG, Barak LS, et al. Atypical responsiveness of the orphan receptor GPR55 to cannabinoid ligands. The Journal of Biological Chemistry. 2009;**284**(43):29817-29827

[75] Pertwee RG. Pharmacological actions of cannabinoids. Handbook of Experimental Pharmacology. 2005;**168**:1-51

[76] Ross RA. The enigmatic pharmacology of GPR55. Trends in Pharmacological Sciences. 2009;**30**(3):156-163

Chapter 10

White Matter Tracts Visualized by Parvalbumin in Nonhuman Primates

Kathleen Rockland

Additional information is available at the end of the chapter

http://dx.doi.org/10.5772/intechopen.70510

Abstract

A well-developed white matter (WM) is one of the characteristics of the primate brain. WM compartments ("tracts" or "bundles") are easily discernible by myelin or neurofilament stains, anterograde tracer injections in nonhuman primates (NHP), and, more recently, diffusion MRI. Relatively overlooked is the fact that several corticofugal and thalamocortical compartments and tracts can be visualized by immunohistochemistry (IHC) for calcium-binding proteins. Since this technique can be easily carried out on postmortem tissues, IHC for calcium-binding proteins is potentially an important bridge for comparisons between NHP and human tissues. This chapter attempts a brief overview of three WM tracts visualized by the calcium-binding protein parvalbumin (PV), as well as a description of the probable origin of the two corticofugal tracts; namely, from PV+ pyramidal cells. Furthermore, the complex, intertwining trajectory of callosal axons is illustrated by single axon reconstruction of five small groups of parietal cortical axons, anterogradely labeled by biotinylated dextran amine.

Keywords: axon trajectory, biotinylated dextran amine, calcium-binding proteins, calbindin-positive pyramidal neurons, corpus callosum, corticopontine tract, geniculocortical tract, parvalbumin-positive pyramidal neurons, single axon reconstruction

1. Introduction

The primate brain exhibits abundant species-specific specializations. Among others, it is gyrencephalic and has an appreciable amount of white matter. These features occur in other large brain species (for example, elephants, cetaceans, and large carnivores; cf.: http://brain-maps.org/) and are not in themselves defining features for primate brains. Nevertheless,

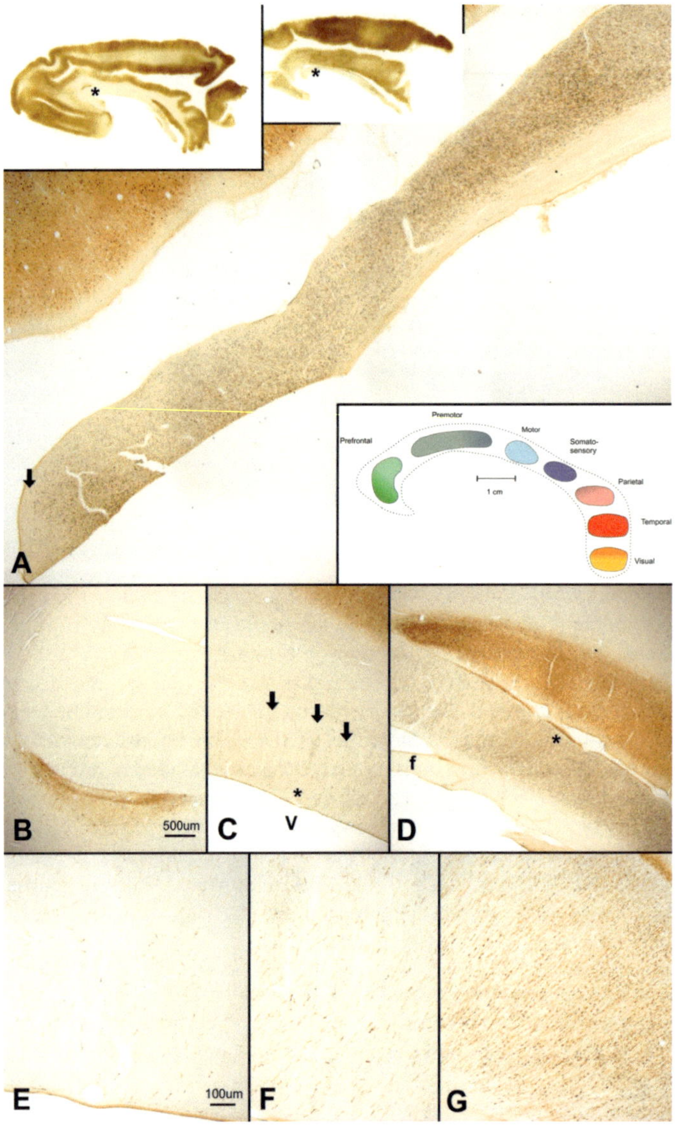

Figure 1. Sagittal sections of the corpus callosum in macaque monkey to illustrate PV+ fibers crossing in the corpus callosum. Anterior is to the left. (**A**) Low magnification overview near the mid-sagittal plane. Part of the anteroventral callosum was trimmed for a separate study, and the section is tilted from the horizontal plane for the sake of formatting. The density of PV+ fibers increases progressively from anterior to posterior. The zone of peak density corresponds to the territory of crossing premotor and motor axons (see schematic at lower right). The short arrow marks a distinct fall-off in density of PV+ fibers, anteriorly, in the vicinity of the rostrum (cut from the section). The histological sections at the upper left show the mediolateral planes of the fields illustrated in **A** (from the section at right), and **B–D** (from the section at left, which is 3.0 mm lateral to the midline). Asterisks = anterior corpus callosum. (**B–D**) Three photomicrographs sampling the anterior (**B**), middle (**C**), and posterior (**D**) regions of the corpus callosum (from the left histological section in **A**. Note increasing density of PV+ fibers with posterior progression. v = ventricle and f = fornix (devoid of PV+ fibers). The arrows in **C** point to three of the organized fascicles of PV+ fibers evident within the larger bundle. Asterisks in **C** and **D** indicate fields at a higher magnification in **F** and **G**. (**E–G**) Higher magnification from anterior (**E**), middle (**F**), and posterior (**G**) regions of the corpus callosum, showing increasing density of PV+ fibers. Scale bar in **B** applies to **A, C,** and **D**. Scale bar in **E** applies to **F, G**. Schematic inset in **A** is reproduced from [12]; Figure 1.

they do usefully set apart primate brains from the lissencephalic and white matter-poor brains of rats and mice. Accordingly, this chapter focuses on several aspects of white matter (WM) organization in macaque monkeys.

The WM in primates has long been known to have suborganization and compartments ("tracts") (as reviewed in [1]). These have traditionally been identified on the basis of origin and target (for example, "corticospinal tract"). WM organization is most frequently investigated from classical myelin stains [2], diffusion MR imaging, and tractography (as reviewed in [3]), or, for nonhuman primates (NHP), labeled profiles after injection of anterograde tracers (as reviewed in [1]). Antibodies against neurofilaments (e.g., SMI32 or SMI312) can also be used [4]. Often overlooked is the fact that some WM tracts can effectively be visualized by immunohistochemistry (IHC) for calcium-binding proteins (calbindin, calretinin, or parvalbumin). These tracts include some callosally projecting axons and some corticofugal projections, originating from parvalbumin-positive (PV+) pyramidal neurons in motor and other cortical areas, as well as thalamocortical projections from PV+ thalamic projection neurons.

Importantly, visualization of neural structures by IHC for PV and the other calcium-binding proteins is a widely applicable technique and, in particular, can be used on immersion-fixed postmortem tissues. This is a definite advantage for morphofunctional investigations of human brains and allows (1) for easier extrapolation between human brains and those of the experimentally more accessible NHPs, (2) for assessment of neural changes between normal and abnormal conditions, and (3) for at least partial validation of tractography imaging data by the histological "gold standard."

This chapter focuses on three PV+ WM compartments (callosal, corticofugal, and thalamocortical) and the distribution of their likely cells of origin (see schematic in **Figure 1**). In addition, data from tracer injections relevant to the trajectory of callosal axons are included.

2. Methods

The data presented here are derived from four macaque brains histologically sectioned and reacted for PV, and two additional macaque brains with injections of the anterograde tracer biotinylated dextran amine (BDA) in inferior parietal cortex [5, 6]. Experimental protocols were all approved by the IACUC at the University of Iowa or the Animal Care Committee at RIKEN Institute (Wako-shi, Japan) and carried out in strict conformance with the NIH Guide for the care and use of laboratory animals (NIH Publication No. 80–23; revised 1996). Every effort was made to minimize the number of animals used and any pain or discomfort experienced by them. As a terminal step, animals were deeply anesthetized with ketamine (11 mg/kg, i.m.) and Nembutal (overdose, 75 mg/kg, i.p.) and were perfused transcardially, in sequence, with 0.9% saline containing 0.5% sodium nitrite, 4% paraformaldehyde in 0.1 M phosphate buffer (PB, 4 L over 30 minutes, pH 7.3), and chilled aliquots of 0.1 M PB with 10, 20, and 30% sucrose.

Brains were removed and, after equilibrating in 30% sucrose buffer, sectioned at 50 μm on a freezing microtome. For PV IHC [7], sections were incubated for 1 h in 0.1 M PB saline (PBS; pH 7.3) containing 0.5% Triton X-100 and 5% normal goat serum (PBS-TG) at room temperature and then for 40–48 h at 4°C with PBS-TG containing mouse monoclonal anti-PV antibody (Swant, Bellinzona, Switzerland; 1:50,000). After rinses, the sections were placed in PBS-TG containing biotinylated goat antimouse IgG (Vector Labs, 1:200) for 1.5 h at room temperature. Immunoreactivity was visualized by ABC incubation (one drop of reagent per 7 ml in 0.1 M PB; ABC Elite kits; Vector Labs) followed by diaminobenzidine (DAB) histochemistry with 0.03% nickel ammonium sulfate.

For the two monkeys with injections of the anterograde tracer BDA, surgery was carried out under sterile conditions after the animals were deeply anesthetized with barbiturate anesthesia (25 mg/kg Nembutal, i.v., following a tranquilizing dose of 11 mg/kg ketamine, i.m.). Parietal cortical areas of interest were localized by direct visualization, subsequent to craniotomy and durotomy, in relation to sulcal landmarks (i.e., the intraparietal and superior temporal sulci). Injections were made by pressure through a Hamilton syringe (10% BDA in 0.0125 M (PBS, 0.5–2.0 μl per injection); Molecular Probes, Eugene, Oregon).

Animals were allowed to recover and survived 18–29 days after injections. They were then reanesthetized, given an overdose of Nembutal (75 mg/kg), and perfused as described above. Brains were cut serially in the coronal plane by frozen microtomy (at 50 μm thickness) and processed histologically for BDA, as described in [5, 6]. Tissue was reacted for 20–24 h in avidin-biotin complex (ABC Elite kits; Vector Laboratories, Burlingame, California) at room temperature (one drop of reagent per 7 ml of 0.1 M PBS). In the final step, BDA was demonstrated by DAB histochemistry with the addition of 0.5% nickel-ammonium sulfate.

Regions of interest (ROIs) were digitized on a Zeiss Axiophot microscope using a 2.5×, 5×, or higher power objective. These high resolution images were then, if needed, stitched using the pairwise stitch function in Image J. Low magnification images of tissue sections were obtained by a high resolution flatbed scanner (Epson Perfection V700 Photo). Figures were assembled in GIMP or powerpoint and saved as JPEG (300 dpi). Axon reconstruction was carried out by camera lucida at low (5× objective) and higher magnifications (20× and 40× objectives).

3. Results

3.1. PV+ callosal axons

A subset of callosal axons in NHP is PV+, and this appears not to be the case for rat or mouse brains (**Figures 1–3**). The PV+ axons are evident through a large anterior-posterior (AP) extent of the callosum, but they are absent or sparse in the most anterior and most posterior portions (**Figure 1**). This corresponds to axons crossing between parts of prefrontal cortex (anteriorly) or early visual cortices (posteriorly), which are evidently PV−.

PV+ callosal fibers were consistently seen in all four brains, although staining in one brain was both fainter and more restricted in the AP extent, possibly suggesting intersubject variability in amount of PV expression. Inspection of online material (brainmaps.org for macaque and see [8] for human brain) confirms the occurrence of PV+ fibers in the WM and specifically in the callosum, although these seemed somewhat less abundant than in our material.

In the coronal plane of section, most fibers are cut in segments of short to intermediate lengths, as would be expected from coronal sectioning of fibers having an overall medial-lateral orientation and trajectory (**Figure 2**). At the light microscopic level, diameters vary from <1.0 to 2.0–4.0 μm. A previous analysis of BDA-labeled callosal fibers reported a range of 0.6–1.2 μm [9]. This study further found that the thickest axons originated from primary motor, somatosensory, and visual areas and the thinnest from prefrontal and temporal areas. Consistent with these findings, another investigation reports the majority of callosal axons as having a diameter of less than 1.0 μm [10], and an electron microscopic study of human and macaque brain also found average values below 1.0 μm [11].

Several papers have investigated fiber diameter, length, and trajectory, in part to evaluate impulse speed and conduction delay of callosal axons, in macaques and humans [10–12]. An early influential paper [13] used electron microscopy and IHC for glia to investigate and compare features of four cerebral commissures in macaque monkey. The increasing use of diffusion MR tractography, alone or in combination with stereology [14], offers an important new tool for WM analysis. The availability of a robust IHC marker, as described here for PV+ fibers, would facilitate comparisons across different age points, different conditions, and different species. Three-dimensional analysis, via serial sections or "clarified" tissue slabs, could easily reveal whether and how often branching occurs in the callosum and support quantification of axon diameters or clustering.

3.2. PV+ corticopontine tract

Other fiber tracts originating from PV+ corticofugal neurons (i.e., corticostriatal, corticorubral, or corticobulbar) are also identifiable by IHC. **Figure 4** documents PV+ corticostriatal and corticopontine fibers, and any number of images on the web (for human, see [8] and for macaque: http://brainmaps.org/index.php?action=viewslides&datid=22, see [4]) will show comparable PV+ tracts. Fibers in the corona radiata, immediately subjacent to motor cortex, contain a particularly dense PV+ population. Many of these are thick (6.0 μm in diameter) and exhibit a distinctive contorted geometry (**Figure 4**). Several surveys of fiber diameters have been published for corticofugal tracts [15–17].

An interesting question is whether the PV+ callosal connections are collateral branches of any of these corticofugal projections. In mice and rats, at least some corticostriatal neurons send collaterals to contralateral cortex [18], and contralateral corticostriatal projections have been described in [19]. Collateral branches from layer 5 neurons to thalamus and brainstem have been discussed in the context of links between perception and action ("efference copy") and may be another instance of primate specializations [20].

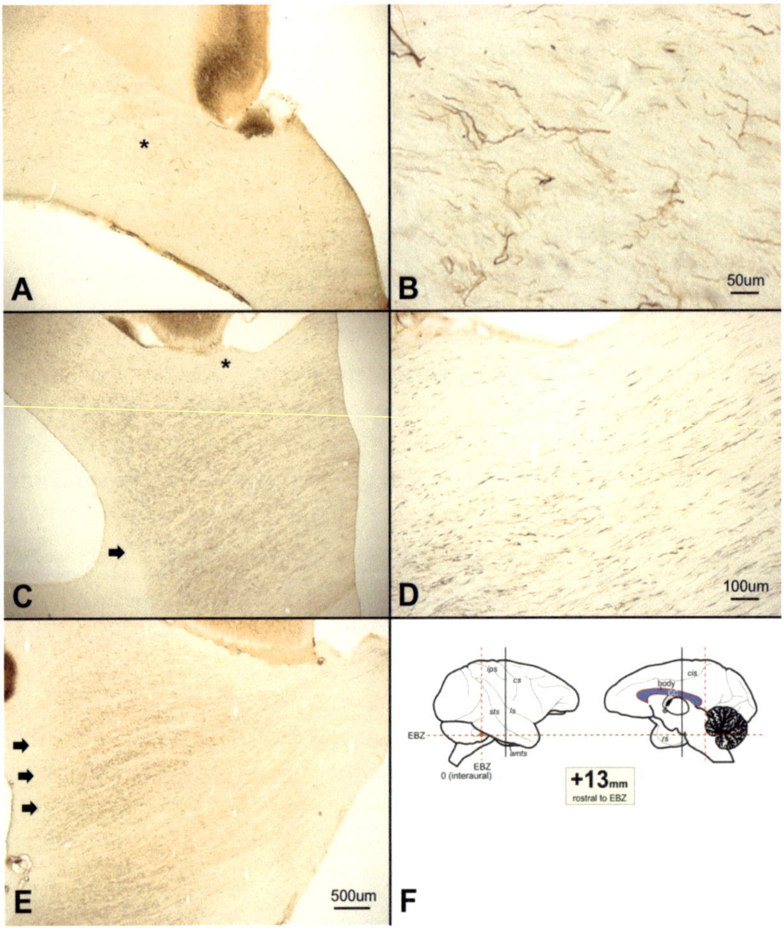

Figure 2. Coronal sections in macaque monkey to illustrate PV+ fibers crossing in the corpus callosum. Midline is to the right. (**A, C, E**) Anterior one-third, middle, and posterior AP levels, where **C** corresponds to the level of the solid vertical line in the right schematic hemisphere in **F**. Arrow in **C** calls attention to the sharp ventrolateral border of PV+ callosal fibers. Arrows in **E** highlight three of the fasciculations within the PV+ tract (see also **Figure 1C**). (**B** and **D**) Higher magnification from the regions of the asterisks in **A** and **C**. Note the scattered large fibers in **B** (and compare with the larger and more numerous large fibers in **Figure 4**). (**F**) Schematic of a right hemisphere, lateral, and medial surfaces (at the left and right of the schematic, respectively). The vertical lines indicate the coronal plane of section, with the more anterior, solid line corresponding to the level in **C**. The more posterior dashed line corresponds to the fields in **Figure 6** (and also the coronal histology section in **Figure 8C**). Reproduced from [4]; **Figure 6**. Scale bar in **E** applies to **A** and **C**.

3.3. PV+ cells of origin

PV is generally associated with subtypes of GABAergic, inhibitory interneurons, mainly basket and chandelier cells. While some GABAergic neurons have been reported to send axons through the corpus callosum [21], the numbers are small, much fewer than the number of PV+ callosal axons illustrated here.

A probable origin of the PV+ callosal fibers is the population of glutamatergic excitatory pyramidal neurons that co-label with PV (**Figures 5** and **6**). These include large Betz cells in layer

http://dx.doi.org/10.5772/intechopen.70510

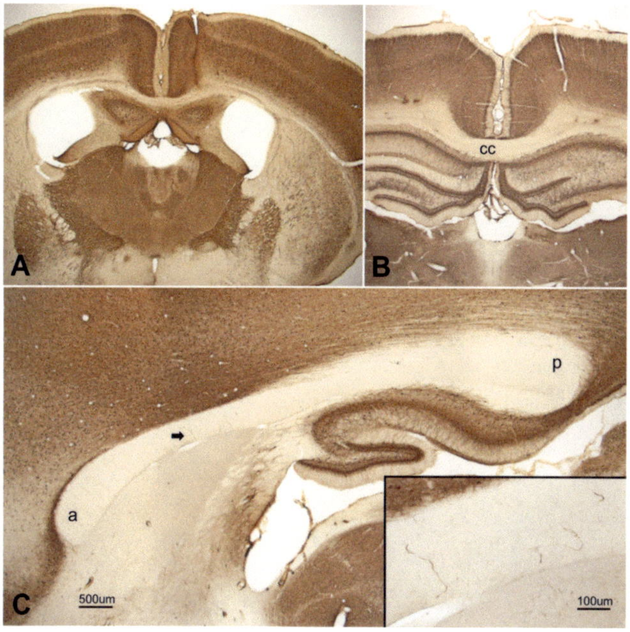

Figure 3. Photomicrographs of (**A**) a coronal section of mouse brain and (**B**) a coronal section of rat brain, both reacted for PV. Note the lack of PV+ fibers at the midline of the corpus callosum ("cc" in **B**). (**C**) Sagittal section through the corpus callosum in rat, slightly lateral to the midline. A few, scattered PV+ fibers are apparent (arrow, and at higher magnification: Inset). a = anterior and p = posterior. Scale bar in **C** applies to **A** and **B**.

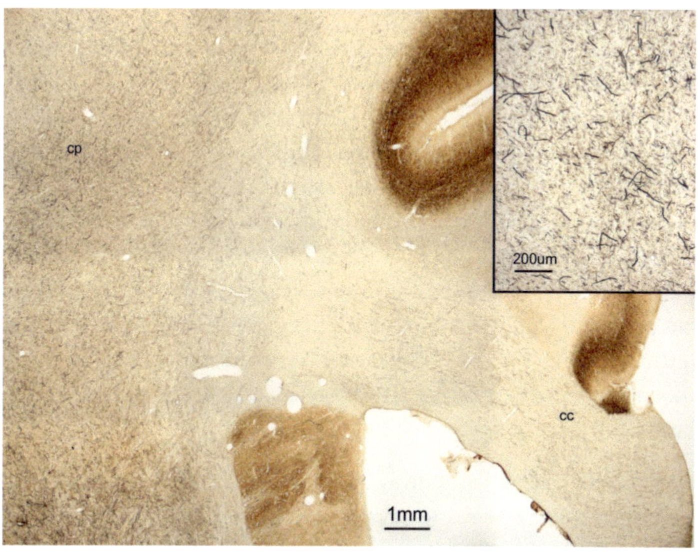

Figure 4. Dorsomedial segment of a coronal section (see solid vertical line in **Figure 2F**) to compare PV+ fibers in the corpus callosum (cc: at right) and those in the lateral white matter (coronal radiata and dorsal part of the internal capsule). The latter are subjacent to overlying motor cortex and partly correspond to corticopontine fibers (CP). CP fibers are thicker and often contorted (see higher magnification inset: upper right).

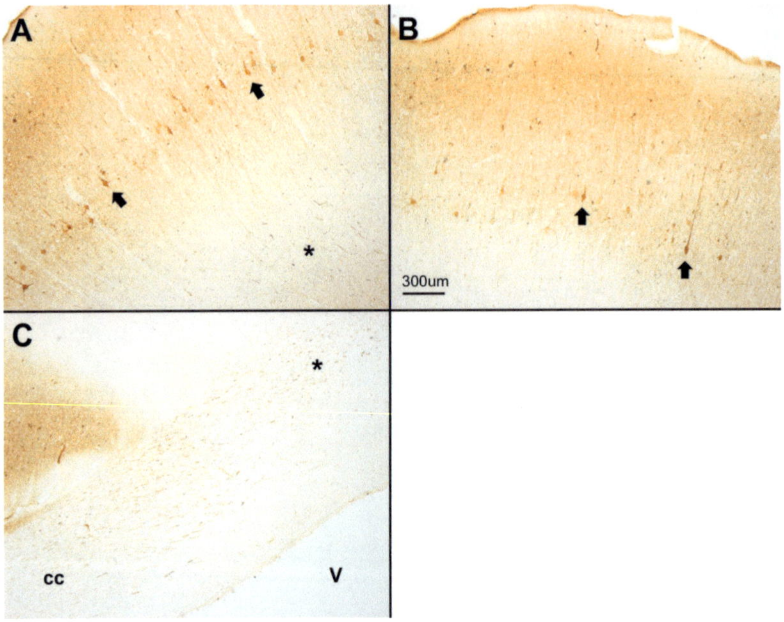

Figure 5. (**A** and **B**) PV+ Betz cells in motor cortex of macaque (coronal sections). Arrows point to two of the PV+ cells in each frame. PV+ interneurons and terminations are also evident. Asterisk in **A** highlights PV+ axons in the underlying white matter. AP level is slightly anterior to the solid vertical line in **Figure 2F**. (**C**) Another example, from the same brain, of PV+ fibers, putatively originating from the PV+ Betz cells in layer 5 and crossing in the corpus callosum (cc). Asterisk marks the callosal tract in a dorsolateral position, as it approaches the midline (at left in **C**). v = ventricle. Scale bar in **B** applies to **A–C**.

5 of motor cortex, large Meynert cells at the border of layers 5 and 6 of primary visual cortex, and large layer 5 pyramidal neurons in several other cortical areas [7, 22]. In histological sections of motor cortex, the PV+ pyramidal neurons typically measure 20 × 50, 30 × 40, or 20 × 30 μm. In the other areas, they can be as large as in motor cortex (i.e., 20 × 50 μm in the posterior cingulate) or toward the smaller range (20 × 30 μm), as in the parietal area 5 (**Figures 5** and **6**). These unusually PV+ pyramidal cells often have a more diffuse DAB filling (i.e., lighter brown) than adjacent PV+ interneurons, perhaps indicative of lower levels of PV (**Figure 6E**). The distribution of these neurons closely parallels that of corticospinal or corticopontine projecting neurons in frontal, premotor, and parietal areas, among others ([23, 24] and **Figure 7**). Thus, the question arises, as noted above, of whether at least some of the callosal connections could be collaterals of corticosubcortical projections originating from PV+ pyramidal cells.

For Betz and Meynert cells, these PV+ neurons also express the Kv3.1b potassium channel, which is usually associated with the fast-firing properties of PV+ interneurons [7, 25]. Physiological studies report that the PV+ pyramidal neurons in NHP exhibit short duration "thin spikes," in contrast to long-duration action potentials, characteristic of neurons in rat motor cortex [26]. The functional significance of this result has been discussed as relating to the fast-conducting property of macaque corticospinal neurons, with the conjecture that fast-conducting corticospinal

http://dx.doi.org/10.5772/intechopen.70510

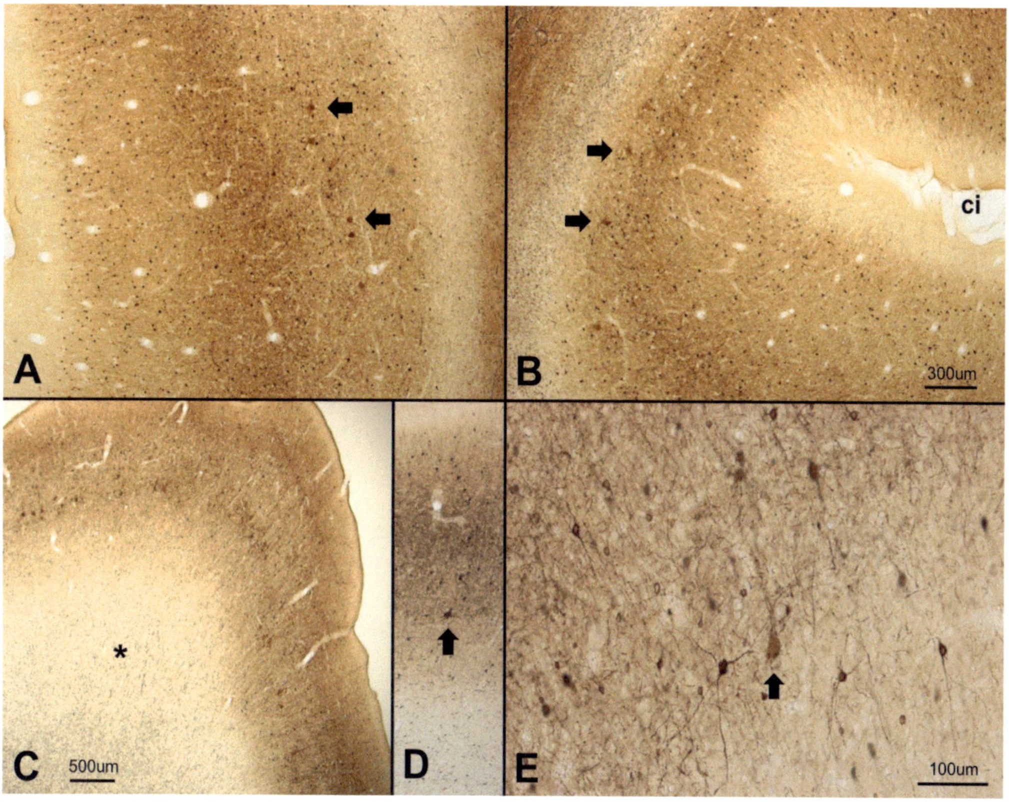

Figure 6. PV+ pyramidal neurons. These are the likely origins of PV+ callosal and/or corticosubcortical fibers. (**A**) A cluster of PV+ pyramidal neurons in layer 5, in the upper bank of the intraparietal sulcus. Arrows point to two of these. Medial is to the right (for **A–C**). (**B**) Arrows point to two of several PV+ pyramidal neurons in layer 5 at the depth of the cingulate sulcus (ci). (**C**) For comparison, a low magnification view from the same brain of PV+ Betz cells in motor cortex (approximately at the vertical dark line in **Figure 2F**). Asterisk marks PV+ fibers in the immediately subjacent white matter, probably corresponding to a mix of outgoing callosal and corticosubcortical fibers. (**D**) PV+ pyramidal neuron (arrow) in parietal area 5. **A, B,** and **D** are at approximately the same AP level as the dashed red line in **Figure 2F** (and see also coronal section in **Figure 8C**). (**E**) Higher magnification image of a PV+ pyramidal neuron (arrow) in a field of PV+ interneurons within the gray matter of the superior parietal lobule. Note lighter intensity of the PV stain for the pyramidal neuron. Scale bar in **B** applies to **A** and **D**.

neurons in macaque may make monosynaptic connections with the motoneurons innervating the most distal muscles controlling the fingers and toes (as discussed in [26]), and thus might be unique to dexterous primate species. This interpretation, however, does not readily apply to PV+ pyramidal neurons in areas outside motor cortex proper.

PV+ excitatory pyramidal neurons have been consistently reported in NHP [27] and, recently, in mice [28]. In PV-Cre transgenic mice, GFP-positive pyramidal neurons were found in layer 5 in a broad swath of cortical areas. In this preparation, the greatest number was in somatosensory cortex, but there were also PV+ pyramidal neurons in motor and visual areas. PV+ corticostriatal pyramidal neurons have been reported by co-labeling with retrograde tracers in mice. These were mainly in layer 5 of retrosplenial and somatosensory areas [29].

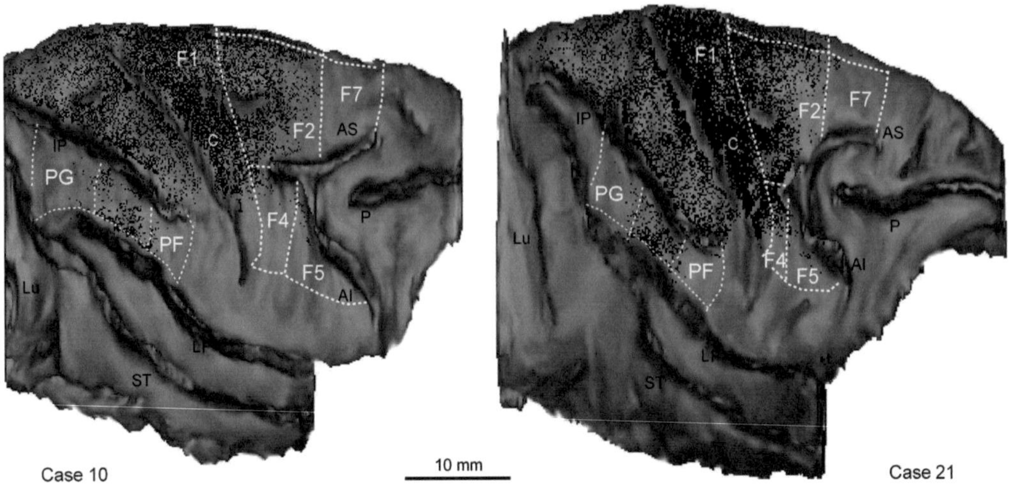

Figure 7. Distribution of retrogradely labeled neurons following retrograde tracer injections in the spinal cord at upper cervical levels (one dot = one neuron). Dorsolateral views of the cerebral hemispheres are shown for two brains (cases 10 and 21) contralateral to the injections. Reproduced from [24]; **Figure 15**.

3.4. Pyramidal cells positive for calbindin or calretinin

As a side observation, we should note that other CBPs are also expressed in subpopulations of glutamatergic cortical neurons, although with pronounced area and species differences [27, 30]. In rodents, there are numerous calbindin + (CB+) pyramidal neurons. CB+ pyramidal neurons are reported in NHP, mainly in the supragranular layers. In NHP, the density of CB+ pyramidal neurons has been described as an increasing gradient from primary visual cortex to temporal association areas [31], but, since a subpopulation of hippocampal CA1 pyramidal neurons is CB+ [32], in both rodents and NHPs, it could also be considered a *decreasing* density from CA1, through temporal association areas, to early and primary visual areas. Calretinin + pyramidal cells are reported in layer 5 of anterior cingulate cortex [27] and in deeper layers of entorhinal cortex in humans [33]. Fine analysis of CB+ or CR+ fiber tracts in NHP is lacking.

3.5. PV+ thalamocortical tract

The thalamocortical tract, including optic radiations (from the lateral geniculate nucleus to primary visual cortex), is a compact, well-delineated bundle in primates and can be readily identified even in classical myelin preparations (e.g., Figure 18-1 in [1]).

In primates, many thalamocortical projections are PV+. An influential distinction has been drawn between thalamocortical terminations that are PV+ and topographically organized vs. those that are CB+ and more diffuse ("core" and "matrix," respectively [34]), although this may hold for only a subdivision of thalamic nuclei. An extensive literature has confirmed that glutamatergic thalamic projection neurons co-localize with either PV or CB, and that their cortical terminations can also be visualized by PV or CB [35, 36]. Relatively under-reported is the fact that the bundles of thalamocortical axons in their WM transit can also be visualized by CBP. This is immediately apparent in whole section atlases or histological images ("sste," external sagittal stratum or optic radiations) in Figures 7 and 17, and level 72b, 77b as in [8] and from Figures 16–23 in the horizontal plane as in [4]. **Figure 8** illustrates PV+ fibers in the thalamocortical tract in macaque (including the optic radiations). An obvious extension of this observation would be double IHC for CB and PV in the same histological preparation.

3.6. Intertwining callosal axons from parietal cortex

Interests in the corpus callosum has burgeoned in this epoch of MRI, and many studies have investigated the topographic organization and axon diameter distribution within the

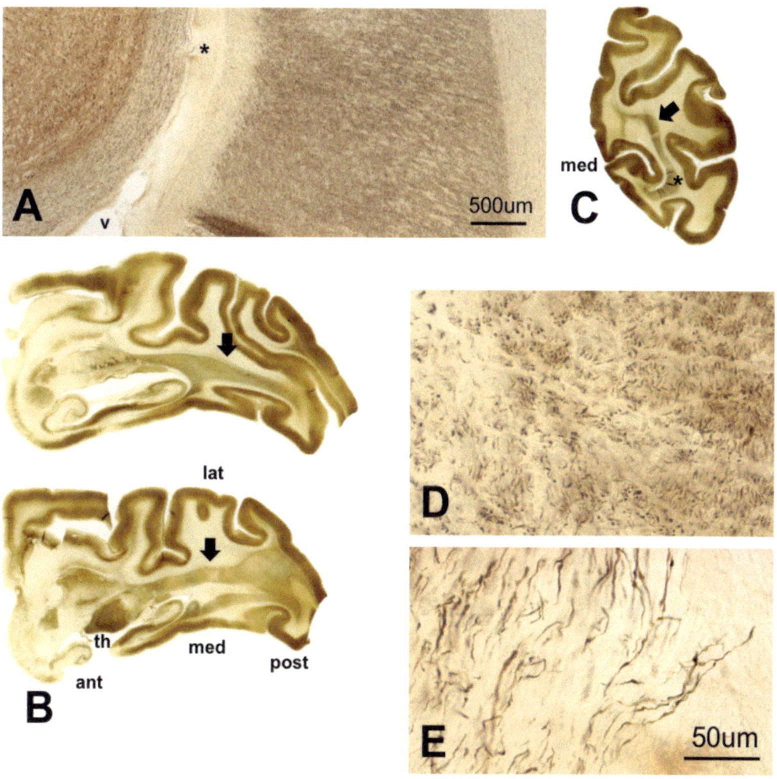

Figure 8. PV+ fibers in the thalamocortical tract. (**A**) Low magnification of the asterisk in **C**, which shows a coronally sectioned histological section. Medial (med) is to the left. Obliquely sectioned geniculocortical axon segments in **A** are subjacent to primary visual cortex. These have exited from the larger, more lateral thalamocortical tract. (**B**) PV+ thalamocortical tract is clearly evident in two horizontally sectioned tissue sections (The upper section is more ventral, and both are below the corpus callosum.) Arrows in **B** and **D** identify the PV+ thalamocortical tract. (**D** and **E**) Higher magnification of the asterisk in **A**, showing detail and different orientation of PV+ fibers. The thalamocortical tract takes a predominantly AP trajectory, so that the coronal plane of section results in fibers cut almost in cross-section. v = ventricle, ant = anterior, lat = lateral, post = posterior, and th = thalamus. Scale bar in **D** applies to **E**.

corpus callosum, both by histological [9–11] and imaging [11, 12] approaches. The trajectory of individual axons, as visualized at high resolution, has been relatively unaddressed. However, serial section analysis (about technique, see [37]) through small segments of the callosum (1.0–2.0 mm AP) reveals that fibers take a complicated, shifting trajectory. **Figure 9** shows five groups of callosal axons (n = 3–5) anterogradely labeled by an injection of BDA in parietal areas (areas 5 or 7) in NHP (see coronal sections at right for orientation). When followed through a reconstructed 3D space (5.4–6.4 mm ML and 2.2–2.8 mm AP), these can be seen to intertwine and exchange dorsoventral position. The "intertwining" (hollow arrows) is within a histological section of 50 μm thickness. At some points (solid arrows), the intertwining is within the same focal plane, compatible with the possibility of physical axon-axon contact.

The principle conclusion from this result is that individual axons do not maintain a stereotyped position but rather shift position within 3D space. An additional conclusion is that closely adjacent axons do not necessarily travel in a parallel trajectory but can intertwine. Similar reconstructions are not available for rodent.

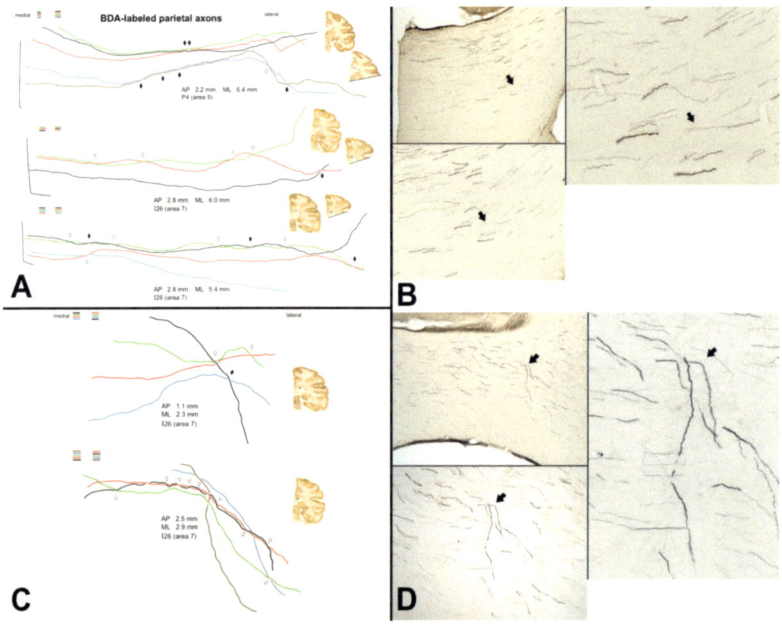

Figure 9. (**A** and **C**) Serial section reconstruction of five groups of parietal callosal axons (consisting of 5, 3, and 4 axons in **A** and 4 and 5 axons in **B**, anterogradely labeled by BDA). As shown for each of the five groups, axons were followed for 1.1–2.8 mm AP and 2.3–6.6 mm ML (mediolateral). Case P4 had a tracer injection in area 5, and case I26 had a large tracer injection in the inferior parietal lobule (area 7). Medial and the cut edge of the corpus callosum are at the left. By color coding the individual axons, it becomes apparent that the order is shuffled, even over these relatively short distances. The medialmost and lateralmost positions of the individual axons are summarized in the color boxes, at upper left for the five axon groups. Small histology sections at the right indicate the AP level of the reconstructions. Intertwinings of individual axons are indicated by solid arrows or, if the crossings occur in the same focal plane, hollow arrows. (**B** and **D**) Histological images corresponding to one coronal histology section (i.e., one AP level) of the axon reconstructions. Arrows point to regions depicted at progressively higher magnification in the three photomicrographs in **B** and **D**.

4. Conclusions

The large, gyrencephalic brains of primates are associated with a well-developed white matter. This has its own organization, in part based on identifiable bundles of axons connecting specific source and target areas. This is increasingly investigated by diffusion MR, although the accuracy of relationships between anatomical substrates and imaging data is still being determined [3, 12, 14]. The white matter also has neurochemically distinct elements. Some of these—the great transmitter-defined glutamatergic, GABAergic, or neuromodulatory fiber tracts—persist across species, although probably with quantifiable species-specific specializations. In primates, as discussed in this chapter, subsystems are characterized by co-localization with PV and other calcium-binding proteins. Callosal and corticofugal PV+ fibers originate from PV+ cell bodies in motor and other cortical areas. PV+ thalamocortical fibers originate from PV+ subpopulations of thalamic neurons. The current interpretation, supported by evidence from the corticospinal tract, is that PV+ excitatory cortical neurons (which are often coincident with neurons positive for the KV3.1b potassium channel) have fast firing properties such as might be needed for fine motor dexterity. Since IHC for PV can be easily carried out in postmortem tissues, this provides a useful morphological label across primate species.

Acknowledgements

I thank Andrew Chang for assistance with figures and manuscript preparation and Dr. Haiyan Gong for the use of her flatbed scanner. Dr. Marie Wintzer (RIKEN Brain Research Institute) contributed substantially to the axon reconstructions in **Figure 9** (Section 3.6).

Author details

Kathleen Rockland

Address all correspondence to: krock@bu.edu

Department of Anatomy and Neurobiology, Boston University School of Medicine, Boston, Massachusetts, United States

References

[1] Schmahmann JD, Pandya DN. Fiber Pathways of the Brain. New York: Oxford University Press; 2006

[2] Dejerine JJ. Anatomie des Centres Nerveux. Paris: Rueff et Cie; 1895

[3] Wandell BA. Clarifying human white matter. Annual Review of Neuroscience. Jul 8, 2016;**39**:103-128

[4] Saleem KS, Logothetis NK. A Combined MRI and Histology Atlas of the Rhesus Monkey Brain in Sterotaxic Coordinates. 2nd ed. San Diego: Elsevier/Academic Press; 2012

[5] Ding S-L, Van Hoesen G, Rockland KS. Inferior parietal lobule projections to the presubiculum and neighboring ventromedial temporal cortical areas. The Journal of Comparative Neurology. 2000 Oct 2;**425**(4):510-530

[6] Zhong Y-M, Rockland KS. Inferior parietal lobule projections to anterior inferotemporal cortex (area TE) in macaque monkey. Cerebral Cortex. 2003 May;**13**(5):527-540

[7] Ichinohe N, Watakabe A, Miyashita T, Yamamori T, Hashikawa T, Rockland KS. A voltage-gated potassium channel, Kv3.1b, is expressed by a subpopulation of large pyramidal neurons in layer 5 of the macaque monkey cortex. Neuroscience. 2004;**129**(1):179-185

[8] Ding S-L, Royall JJ, Sunkin SM, Ng L, Facer BAC, Lesnar P, et al. Comprehensive cellular-resolution atlas of the adult human brain. The Journal of Comparative Neurology. Nov 1, 2016;**524**(16):3127-3481

[9] Tomasi S, Caminiti R, Innocenti GM. Areal differences in diameter and length of corticofugal projections. Cerebral Cortex. 2012 Jun;**22**(6):1463-1472

[10] Phillips KA, Stimpson CD, Smaers JB, Raghanti MA, Jacobs B, Popratiloff A, et al. The corpus callosum in primates: Processing speed of axons and the evolution of hemispheric asymmetry. Proceedings of the Biological Sciences. Nov 7, 2015;**282**(1818):20151535

[11] Liewald D, Miller R, Logothetis N, Wagner H-J, Schüz A. Distribution of axon diameters in cortical white matter: An electron-microscopic study on three human brains and a macaque. Biological Cybernetics. Oct 2014;**108**(5):541-557

[12] Caminiti R, Carducci F, Piervincenzi C, Battaglia-Mayer A, Confalone G, Visco-Comandini F, et al. Diameter, length, speed, and conduction delay of callosal axons in macaque monkeys and humans: Comparing data from histology and magnetic resonance imaging diffusion tractography. The Journal of Neuroscience. Sep 4, 2013;**33**(36):14501-14511

[13] Lamantia AS, Rakic P. Cytological and quantitative characteristics of four cerebral commissures in the rhesus monkey. The Journal of Comparative Neurology. Jan 22, 1990;**291**(4): 520-537

[14] Charvet CJ, Hof PR, Raghanti MA, Van Der Kouwe AJ, Sherwood CC, Takahashi E. Combining diffusion magnetic resonance tractography with stereology highlights increased cross-cortical integration in primates. The Journal of Comparative Neurology. Apr 1, 2017;**525**(5):1075-1093

[15] Morecraft RJ, McNeal DW, Stilwell-Morecraft KS, Dvanajscak Z, Ge J, Schneider P. Localization of arm representation in the cerebral peduncle of the non-human primate. The Journal of Comparative Neurology. Sep 10, 2007;**504**(2):149-167

[16] Firmin L, Field P, Maier MA, Kraskov A, Kirkwood PA, Nakajima K, et al. Axon diameters and conduction velocities in the macaque pyramidal tract. Journal of Neurophysiology. Sep 15, 2014;**112**(6):1229-1240

[17] Innocenti GM, Vercelli A, Caminiti R. The diameter of cortical axons depends both on the area of origin and target. Cerebral Cortex. Aug 2014;**24**(8):2178-2188

[18] Shepherd GMG. Corticostriatal connectivity and its role in disease. Nature Reviews. Neuroscience. Apr 2013;**14**(4):278-291

[19] Innocenti GM, Dyrby TB, Andersen KW, Rouiller EM, Caminiti R. The crossed projection to the striatum in two species of monkey and in humans: Behavioral and evolutionary significance. Cerebral Cortex. Jun 9, 2016:3217-3230. DOI: 10.1093/cercor/bhw161

[20] Guillery RW. Anatomical pathways that link perception and action. Progress in Brain Research. 2005;**149**:235-256

[21] Peters A, Payne BR, Josephson K. Transcallosal non-pyramidal cell projections from visual cortex in the cat. The Journal of Comparative Neurology. Dec 1, 1990;**302**(1):124-142

[22] Preuss TM, Kaas JH. Parvalbumin-like immunoreactivity of layer V pyramidal cells in the motor and somatosensory cortex of adult primates. Brain Research. Mar 18, 1996; **712**(2):353-357

[23] Geyer S, Matelli M, Luppino G, Zilles K. Functional neuroanatomy of the primate isocortical motor system. Anatomy and Embryology. Dec 2000;**202**(6):443-474

[24] Rozzi S, Calzavara R, Belmalih A, Borra E, Gregoriou GG, Matelli M, et al. Cortical connections of the inferior parietal cortical convexity of the macaque monkey. Cerebral Cortex. Oct 2006;**16**(10):1389-1417

[25] Constantinople CM, Disney AA, Maffie J, Rudy B, Hawken MJ. Quantitative analysis of neurons with Kv3 potassium channel subunits, Kv3.1b and Kv3.2, in macaque primary visual cortex. The Journal of Comparative Neurology. Oct 1, 2009;**516**(4):291-311

[26] Soares D, Goldrick I, Lemon RN, Kraskov A, Greensmith L, Kalmar B. Expression of Kv3.1b potassium channel is widespread in macaque motor cortex pyramidal cells: A histological comparison between rat and macaque. The Journal of Comparative Neurology. Feb 18, 2017:2164-2176. DOI: 10.1002/cne.24192

[27] Hof PR, Glezer II, Condé F, Flagg RA, Rubin MB, Nimchinsky EA, et al. Cellular distribution of the calcium-binding proteins parvalbumin, calbindin, and calretinin in the neocortex of mammals: Phylogenetic and developmental patterns. Journal of Chemical Neuroanatomy. 1999 Feb;**16**(2):77-116

[28] Tanahira C, Higo S, Watanabe K, Tomioka R, Ebihara S, Kaneko T, et al. Parvalbumin neurons in the forebrain as revealed by parvalbumin-Cre transgenic mice. Neuroscience Research. Mar 2009;**63**(3):213-223

[29] Jinno S, Kosaka T. Parvalbumin is expressed in glutamatergic and GABAergic corticostriatal pathway in mice. The Journal of Comparative Neurology. Sep 13, 2004;**477**(2):188-201

[30] Suzuki WA, Porteros A. Distribution of calbindin D-28k in the entorhinal, perirhinal, and parahippocampal cortices of the macaque monkey. The Journal of Comparative Neurology. Sep 30, 2002;**451**(4):392-412

[31] Kondo H, Tanaka K, Hashikawa T, Jones EG. Neurochemical gradients along monkey sensory cortical pathways: Calbindin-immunoreactive pyramidal neurons in layers II and III. The European Journal of Neuroscience. Dec 1999;**11**(12):4197-4203

[32] Mizuseki K, Diba K, Pastalkova E, Buzsáki G. Hippocampal CA1 pyramidal cells form functionally distinct sublayers. Nature Neuroscience. Aug 7, 2011;**14**(9):1174-1181

[33] Mikkonen M, Soininen H, Pitkänen A. Distribution of parvalbumin-, calretinin-, and calbindin-D28k-immunoreactive neurons and fibers in the human entorhinal cortex. The Journal of Comparative Neurology. Nov 10, 1997;**388**(1):64-88

[34] Jones EG. Viewpoint: The core and matrix of thalamic organization. Neuroscience. Jul 1998;**85**(2):331-345

[35] DeFelipe J, Jones EG. Parvalbumin immunoreactivity reveals layer IV of monkey cerebral cortex as a mosaic of microzones of thalamic afferent terminations. Brain Research. Oct 18, 1991;**562**(1):39-47

[36] Melchitzky DS, Sesack SR, Lewis DA. Parvalbumin-immunoreactive axon terminals in macaque monkey and human prefrontal cortex: Laminar, regional, and target specificity of type I and type II synapses. The Journal of Comparative Neurology. May 24, 1999;**408**(1):11-22

[37] Rockland KS. Visual cortical organization at the single axon level: A beginning. Neuroscience Research. Mar 2002;**42**(3):155-166